HUMANLY POSSIBLE

HUMANLY POSSIBLE

A Biologist's Notes on the
Future of Mankind

JEAN ROSTAND

Translated by Lowell Bair

Saturday Review Press
New York

First published in France under the title *Le Courrier d'un Biologiste*

Copyright © 1970 by Jean Rostand and Editions Gallimard
Translation copyright © 1973 by Saturday Review Press
FIRST AMERICAN EDITION 1973

Published simultaneously in Canada by Doubleday Canada Ltd., Toronto.

Library of Congress Catalog Card Number: 72-79038
ISBN 0-8415-0177-7

Saturday Review Press
230 Park Avenue
New York, New York 10017

PRINTED IN THE UNITED STATES OF AMERICA
Design by Margaret F. Plympton

CONTENTS

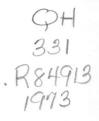

HUMANLY POSSIBLE

A BIOLOGIST'S MAIL

In an earlier book I mentioned the strange letters that a biologist sometimes finds in his mail. Strange because of the wishes they express, the questions they ask, the illusions they show, the reproaches they contain. I now want to return to that subject, extending it to cover all the letters that come to me, some of which are only indirectly connected with biology. They will give an idea of the relations that can arise between the public and a solitary man who, having written a great deal—perhaps too much—over the past half-century, has sometimes stirred up sharp reactions by the judgments he has pronounced on a wide range of issues, not only scientific, but also philosophical, social, and moral. Discussing my own reactions to some of these letters will give me a new opportunity to express myself on matters that I feel strongly about.

In the first place, quite naturally, I receive many letters about problems of heredity. I am asked if a certain disease or defect is transmitted hereditarily, and, if so, how. Such questions are usually prompted by an intention to marry, the letter having been written by the prospective bride or groom, or by one of their parents.

The answer is easy to give if it concerns an unequivocally hereditary disease or defect, either dominant or recessive. When a young woman asks whether there will be a risk to her children if she marries a man with brittle bones, she can be told without hesitation that she will have an even chance of giving birth to a child with brittle bones. But there are much less clear-cut cases: maladies with a complex heredity that depends on multiple genes, or morbid predispositions, which, though conditioned by heredity, develop their bad effects only under certain environmental circumstances.

Sometimes a letter not only requests information, it also appeals for advice. "Your letter will be decisive," I am told. "I know there is a certain risk in the marriage I am considering, but do you think that risk is acceptable?"

This raises a great difficulty. How am I to evaluate the acceptability of a risk? It depends on so many things: the individual's temperament, the extent to which he is anxious or hesitant, his economic situation, etc.

Should a young woman with a congenital cataract be advised against marriage? Or one who has two hunchbacked grandfathers? Or a young man afflicted with deafness? Should a woman who has already had three stillborn children be dissuaded from another pregnancy?

Among those who consult me, there are some whose letters show that inwardly they have already made their decision and simply want to be relieved of the responsibility for it. A few go so far as to say, "If your opinion is not favorable, do not answer me. I can always tell myself that my letter must not have reached you."

Some will not accept a rather indefinite answer that is not reassuring enough to suit them. They insistently demand some kind of certainty. With all the progress that science has made, they ask, isn't it possible to examine someone's blood or cells to make sure he can transmit no undesirable traits? Or isn't it possible to prevent a child from having a certain defect? They would gladly undergo all the necessary examinations and tests. And they are so accustomed to the miracles of biology that they are surprised and disappointed to learn that it cannot give them satisfaction in these matters.

There is really only one case in which examination of the cells—or, more precisely, the complements of chromosomes, or karyotypes—can lead to a useful prediction: the case of the parents of a mongoloid child, because one of them, though normal in appearance, is sometimes the carrier of a certain chromosome abnormality that greatly increases the probability of procreating mongoloids.

Fear of the hazards involved in marriage between blood relatives is a source of many letters to me: "I am in love with my first cousin. Are there genetic reasons why we should not marry?" Here again, giving an answer poses certain problems. Children of consanguine parents unquestionably run a slight additional risk, but its nature cannot easily be explained to laymen. The way in which it is presented to them necessarily influences their decision to some degree; and I admit that for my part I am rather inclined to present it in such a way as not to make it seem too alarming.

At any rate, there are so many risks of all kinds in marriage that, it seems to me, those added by consanguinity should not be regarded as prohibitive. By my relative optimism on the subject I have probably helped to bring about a good number of consanguine marriages. Let us hope that their results have not been too bad.

It is of course better, genetically, not to love one's cousin. But if one does, must that love be sacrificed because of eugenic misgivings? A consanguine marriage of love may be better than a nonconsanguine marriage without love, even for the children.

There are also many questions in my mail about the possible incompatibility of parental Rh factors and the transmission of racial characteristics. (Can two light-skinned mulattoes have a child darker than either of them? After how many generations can one be sure that Negroid traits will not emerge?)

And letters concerning the determination of paternity are especially numerous. It is well known that genetic information often makes it possible to declare, not that any given man is the father of any given child, but that a certain man cannot be the father of a certain child. These "exclusions of paternity" are made by means of blood tests. It is not rare for a correspondent to send me the results of such tests and ask me to state what conclusions follow from them.

It is always dangerous to answer these requests. Great caution is advisable in dealing with a letter that asks—simply out of scientific curiosity, the writer says—whether a blue-eyed man and a blue-eyed woman can have a brown-eyed child.

Since the chances are great that such a question is not motivated by a pure desire for knowledge, I am careful, if I answer it, not to be too categorical. I always leave a door open. Do both the man and the woman have really blue eyes? Are there no traces of gray or green? Furthermore, it is always possible for a mutation to occur, breaking the regularity of Mendel's Laws. And so, without betraying scientific truth in any way, I may be preserving the peace of a household.

Telegony, the old and still tenacious superstition of the first father's supposed influence on a woman's later children by another man, shows its persistence in a biologist's mail.

A remarried woman asks if her second husband really has any justification for tormenting himself because he thinks his son has the bushy eyebrows of her first husband.

This superstition goes so far as to arouse the suspicion that a wife may have some genetic influence on her husband. Several women have asked me quite seriously if a remarried widower can transmit any of his first wife's physical or moral traits to the children of his second marriage.

An unusually long pregnancy may cause doubt or anxiety about the legitimacy of the child. A man whose wife has given birth 332 days after being separated from him asks if there is still some chance that he may be the father of that child who has arrived so late.

Sometimes a woman who has taken some sort of medication during the first weeks of pregnancy recalls the thalidomide tragedy and sends me a letter expressing fears that are often groundless. I am usually given no news of the child whose mother was alarmed, but once a woman I had confidently reassured wrote me this charming letter, which I can quote literally without indiscretion:

"Dear Sir: I am happier than anyone else to announce to you that I have achieved biological independence.

"The starting handicap has apparently not hindered my development. Alert and smiling at my daydreams, I would make fun of Mama for having bothered you several months ago if I did not know the anxiety that was gripping her heart at that time. Praise be to Mother Nature for having calmly continued on her way!

"Life has given me a royal gift, for which I am greatly indebted. Thank you, Sir, for having consented to be the guardian of our hope."

Before adopting a child, the parents sometimes ask whether there are scientific means of predicting, to some extent, its physical and moral destiny. This kind of question has naturally become more frequent since the mass media revealed the existence of the chromosome abnormality (doubling of the Y chromosome) that produces a predisposition toward criminal behavior.

I have already referred to this kind of thinking in the mind of the general public. Misled by sensationalistic articles or science fiction stories, many people tend to assume that nothing is now impossible for science, particularly biology. As a result, I receive some astonishing letters that are also quite touching in their naïveté.

Parents who have lost an adored child would like to be given a means—or at least a greater chance—of having another child exactly like the first one.

A mother—a peasant woman—asks whether science, which does so many things that approach the miraculous, could bring her little boy who was killed in an accident back to life. Another, whose child is hopelessly ill, asks whether she should have him embalmed immediately after his death, because she once read in a very serious magazine that, according to a famous professor in California, it will eventually be possible to bring mummified pharaohs back to life from molecules contained in their cells.

The freezing of the human body has recently given rise to hopeful letters. Cells, tissues, and even small organs can now be kept alive indefinitely by freezing them at very low temperatures, so that all vital activity is suspended in them. This well-established fact has led to the conjecture that human organisms can some day be preserved intact—as in Edmond About's novel *L'Homme à l'oreille cassée.*

It cannot be done at present, because the freezing of an entire organism causes irreversible damage. But the American physicist, Robert Ettinger, has written an entire book, *Prospect of Immortality*, around the following idea: since in a century or two, perhaps even sooner, science will be able to repair the damage done by freezing, it is reasonable to freeze dead people now, so that they can later be brought back to life when medicine has become capable of curing the disease that caused their death. This would be a kind of bet on the future, based on faith in the omnipotence of science.

Ettinger's book, halfway between real science and science fiction, is well informed, well written, and full of zesty American humor. I wrote a preface to its French translation without foreseeing the flood of letters that those few paragraphs would bring me, especially after the press announced that Salvador Dali had decided to make both his genius and his mustache immortal by freezing.

Some of my correspondents do not even want to wait for death before having themselves preserved: they are ready to be frozen alive, with the understanding that they will be revived when interplanetary travel has become commonplace. Others, more altruistic, are thinking of freezing an aged or seriously ill relative. Isn't it their duty, they ask, to give him a chance of resurrection, no matter how slight it may be? My answer is unambiguous, and negative. Although I presume that science will some day advance to the point where human beings can be preserved, I do not think that we have anything to gain by premature freezings.

Preservation of the human body by freezing is still only a dream, but preservation of sperm cells is now an established fact. A number of babies conceived from semen frozen for several weeks, or even months, have been born in America. They are perfectly normal children, and there is every reason to believe that they would have been the same if the semen had been kept frozen for decades or centuries. That is the basis of an unusual plan presented to me in a registered letter from a middle-aged gentleman. He would like to have some of his sperm cells frozen now, to be used fifty years after his death for posthumous procreation with a woman having certain strictly specified physical and mental characteristics: she should be tall, beautiful, preferably a blonde with brown eyes, intelligent, cultivated, and reasonably well educated. A sum of money would be deposited with a notary and insured against the risks of inflation to compensate the future mother and provide for the first needs of the child.

The publicity given to transplants—and especially, in recent years, to heart transplants—brings me some strange requests. A man despairing over an unhappy love affair would like to give away his heart immediately—but to another man, not a woman. A man who is in excellent health, but who feels that there is nothing left for him in this world, wants to make a gift of his lungs. An old woman who still has good eyes places them at the disposal of a young blind girl.

Although brain transplants are at present impossible, the idea of them is exciting and gives rise to rather diabolical hopes.

A very old man who, he tells me, has been prevented by circumstances from realizing his true capabilities, and who has vast plans of universal interest, feels that it would be a loss to mankind if his intelligence and cultural advantages were soon reduced to nothing; would it not be possible for him to give his brain to a young man with an irreparable cerebral lesion?

Here we have, essentially, the great dream of Faust. It would indeed be wonderful for an old man if he could place his mind in a young, healthy body, combining the knowledge and experience of age with the vigor of youth.

Such a transplant—unlike all others, including heart transplants—would obviously be to the benefit of the donor, since the brain is the seat of the self, of memory, and personality.

By occupying the body of the "recipient," the "donor" would renew and prolong his existence.

But, as I have said, brain transplants are not possible now. And they probably never will be. Fortunately. For once, we can be glad that nature imposes limits on scientific boldness.

There are also letters in my mail from those who, by any procedure, by any treatment, no matter how risky or even dangerous, would like to recover a little of their youth. A few years, a few months, a few weeks. I believe they would be content with even a few days of rejuvenation.

Then there are those who are not satisfied with their sex. Whenever a newspaper announces that someone has undergone a change of sex, I can be sure that within a week I will receive several letters expressing a desire for a similar metamorphosis. They are usually from men.

One man tells me that he knows he has been a woman, or rather a girl, since earliest childhood; his masculine appearance, which is not even very strongly accentuated, is only a mistake of nature that ought to be corrected, now that it can be done by surgery and the use of hormones. He demands help in "pleading the just cause of his femininity." If necessary, he will address himself to the League of Human Rights to support his claim. And he adds, not without a certain versifying ability:

> Anything can be bought today
> If your wallet is not too slender:
> Health, or beauty, or youth, or even
> A welcome change of gender.

On the subject of sex, I will mention the curious attitude of a man who would like science to make women oviparous, so that development of the fetus outside the mother's body would be substituted for the traditional intrauterine pregnancy: jealous of the maternal abdomen, he feels it would be more equitable if the child were formed in "neutral territory."

As for parthenogenesis, or reproduction by the female without the participation of the male, it is naturally a common theme of my correspondents, since I have studied this question a great deal in toads and frogs and have written several books on it.

Here, the tone of my mail varies greatly: often animated and impassioned, it goes from enthusiasm to furious condemnation. Some women, believing that this type of reproduction is already possible in human beings, offer to try it immediately; others, better informed, volunteer to serve as guinea pigs for experiments in the near future.

These impatient women see a moral victory in parthenogenesis, a decisive conquest for their sex—something like a "decolonization."

"You have no idea," writes one of them, "how many young and beautiful women would like to keep themselves pure and intact. Bestial desire is the opposite of love. Motherhood should not require the sacrifice of virginity."

Another regards parthenogenesis as "a wonderful, incomparable instrument of liberation and emancipation"; it will give the world the chance of a new race, free of the primordial blemish and capable of leading mankind to its supreme destiny.

It is interesting to note that this fantasy of solitary procreation has occurred to exceptional minds. The great poet Anna de Noailles, for example, in her still unpublished *Journal de jeune fille* (1893–1894) writes that she asked God for a child born only of herself: "I wanted another little Anna who would comfort me and understand me." (*Revue de Paris*, January 1956.)

Some women, on the other hand, aggressively protest

against reproduction without the male: "Incompatible with feminine dignity and honor, it would reduce us to the level of animals. Only a degenerate scientist like you could even consider such horrors. There are males and females even among flowers; that is God's plan, and all your abominable science will be powerless to change it. Parthenogenesis is a crime against both the mother and the child. It is an invention of the devil that foreshadows the ruin of the planet announced in the Book of Revelations. Fatherless children may have a human appearance, but they will have no souls. Your only volunteers, Mr. Biologist, will be abnormal, neurotic, and repressed women. Are you so dissatisfied with your own birth? Is that why you have had to concoct such an unnatural form of motherhood?"

But of course, as is to be expected, the most vehement protests come from men. Cut to the quick because of their "castration complex," humiliated for their virility, some men give vent to their anger by insulting me: "You should at least have the decency to keep quiet about your perverse and disgusting research. You want to eliminate the male and abolish love: this is not surprising for someone with a skull as bald as yours. The greatest discovery you could make would be to realize that you are crazy."

An unexpected note of approval in the midst of this hostile chorus: "I am in favor of parthenogenesis, because it will allow a very ugly woman to have children despite the cruelty of the men who have rejected her because of her ugliness. It will give her a kind of revenge against them."

As I have said many times, I am by no means an advocate of human parthenogenesis. I have even pointed out some of its biological drawbacks by showing, in the case of the toad, that virginal procreation produces a relatively high number of abnormal individuals. I have simply said—because it is true— that in all likelihood parthenogenesis will some day be possible for women and that in theory we can conceive of a human race composed entirely of women, while we cannot conceive of one composed entirely of men.

About thirty years ago, Emminarie Jones, a young German woman living in England, claimed to have given birth after a pregnancy that could not be explained by the usual cause. The case created a great stir in the press, and famous biologists, such as Haldane, took an interest in it.

At about the same time I received a letter from a French girl who said that she had also been a victim of spontaneous parthenogenesis. No one would believe in her virginal conception. Her fiancé had abandoned her, and this had plunged her into a severe nervous depression that made her wonder whether, after all, she might have conceived like everyone else. But when she learned of what had happened to the young German woman, she returned to her first conviction: she, too, was a virgin mother.

To finish up the subject of parthenogenesis, I will report that one of my correspondents had urged me to experiment with sheep because he suspects that this type of reproduction occurs naturally with them: if John the Baptist gave Christ the name Lamb of God, it is not only because lambs are very gentle creatures but also because they are sometimes engendered by the action of the Holy Ghost.

It will come as no surprise to learn that I receive many letters concerning frogs. I am asked for scientific information about their anatomy, physiology, behavior, and feeding; the conditions under which they should be raised; the best design for a froggery. But I am also assailed by questions arising from a less scientific kind of curiosity. How far can they jump? Can they be tamed? Do your frogs recognize you? Do you have favorites among them? Do you give each of them a name? Do they have a rudimentary language? Is it true that they can be used as barometers? What method do frog swallowers use?

What is the meaning of a dream about frogs or toads? Do you yourself often dream of frogs? Do you eat them after you have dissected them? I would not be surprised if I were asked for my recipes.

I must say that the frog is commonly used by those of my correspondents who want to discredit one of my judgments on some subject or other. "The fact that you are an expert on frogs does not entitle you to be dogmatic about everything else. Frogs do not hold the answers to all questions. Please spare us your opinions on anything higher than the frog." When I once took the liberty of stating that Teilhard de Chardin was not the great philosopher he was sometimes said to be, I drew this reply: "Sinanthropus has at least as much to teach us as the frog."

Or: "We are willing to listen to you when you talk about frogs, but when it comes to human affairs we prefer other authorities."

My critics never cease harping on the frog. Everything about me is "frogified." I have been told that my long association with frogs has finally made me look like one.

Must I repeat here that the study of frogs has nothing to do with any of my philosophical, political, or moral opinions, which may legitimately be held against me? The frog makes no claim to being able to teach us anything about the human condition, the meaning of life, the future of society, or the different forms of government. I freely admit that the scientific spirit, the biological spirit (whether concerned with frogs or not), and a long familiarity with the phenomena of life may result in a tendency to think along certain lines; but frog researchers come in all varieties: materialists and spiritualists, believers and atheists. As for politics, the frog has never influenced anyone's vote.

I will therefore ask those who disagree with me (as they have a perfect right to do) to leave "my frogs" in peace; they are completely innocent of what I am.

People write to me rather often to express surprise that I spend so much time on frogs. "I don't mean to offend you, but aren't there more important subjects? Aren't there other kinds of research, which can be more profitable to the human race?"

This is not the place to plead for the frog (as I have done elsewhere); I will simply point out that all the problems of biology, great and small, can be studied in that humble animal. It offers inexhaustible material for the investigator's patience, astuteness, inventiveness, and skill. To him it is something like clay to a sculptor, canvas to a painter, or blank paper to a writer. So we will never finish studying the frog. If we knew everything about the frog, we would know everything about life, including human life.

Need I add that if I have once again stressed the importance of the frog, it is not out of personal vanity? The glory of the frog is in no way reflected on me. I am only one of the countless researchers who, since the illustrious Lazzaro Spallanzani, have wrestled with the mysteries of the amphibians. "My frogs," as my correspondents too often refer to them, amount to very little; but the frog is a vast domain.

While there are many who choose to belittle my frogs for a wide variety of reasons, there are also a few who feel called upon to defend them against me.

"Frog butcher! Toad slaughterer!" Yes, I occasionally find that sort of thing in my mail. One lady, for example, sends me a photograph of myself, cut out of a newspaper, in which I am shown handling a frog, and she adds this gracious comment: "If you return to earth—and I firmly believe in the transmigration of souls—I hope it will be in the form of a frog. Justice will then be done: it will be your turn to spread your webbed feet in terror, while a gigantic torturer looks down at you."

The truth is that I have never practiced vivisection on frogs, at least not without anesthesia; and I would like to be sure that these kindly defenders of animals are genuine disciples of Dr. Schweitzer and do not belong to that category of people who indignantly protest against the mistreatment of an animal in a laboratory but placidly accept the prospect of atomic genocide.

There is a tendency to imagine that a biologist must know everything about biology, and even about many other things. This accounts for the extraordinary and bizarre diversity of some of the questions addressed to me.

Will science soon make an artificial man? Why try to prolong life, since electrons are already eternal? Is it true that the atoms of Plato and Henry VIII are still scattered around us? If so, couldn't we devise refined methods of analysis that would enable us to identify them? Do two twins necessarily have the same father? Can a woman with a pointed head still have good-looking children? Are illegitimate children predisposed to have other illegitimate children when they grow up? Can an emotional shock cause pregnancy? How can you tell the sex of a cockroach? Why are there only four human races? Shouldn't lobsters be put to sleep with ether before they are boiled alive? Could frog eggs be used to make caviar? If a Martian came to Earth in a flying saucer, could he have a child by one of our women? Is it true that in America they have very expensive medicines that make people more intelligent? Is there a chromosome that makes biologists, as there is one that makes criminals? Can a young monkey be taught to bowl? Is everything inside a snail's shell edible? Is it because of nuclear explosions that people are becoming more stupid with each generation? Can it be that "spots before the eyes" are chromosomes of the dead? Is it possible that each of us might never have existed? Will we some day know everything about everything? Is euthanasia acceptable for an old dog? Can mouth-to-mouth resuscitation be used on a cat? How do you cure constipation in a turtle?

Extraordinary things are sometimes reported to me in letters. Some of them are well-attested facts: an albino tadpole, a blue tree frog, a one-eyed snail. But in most cases they are illusory, or at least unverifiable: an old man who has grown four sets of teeth, the last one at the age of eighty-eight; an enormous horned toad, covered with fur, that was glimpsed at the edge of a forest and escaped all pursuit.

And then, of course, there are letters that tell me of astonishing inventions. One correspondent has devised a way of accumulating a marvelous fluid he has discovered, a "new energy" that acts on all living beings. Another prepares a kind of magic water that makes tomatoes grow to the size of a small pumpkin and can cause a sow to produce short-legged piglets. Another, after forty years of research, has created several insects, small but quite mobile, by means of special vibrations. Another waits for me every day, including Sundays, between nine and one o'clock, to show me the tiny crab that causes cancer. Another knows how to turn any baby into a great man; but he will explain his discovery only to a government (he has not yet decided which one). Another has quite simply reduced love to a set of formulas. Another takes strolls in the fourth dimension, and he invites me to accompany him whenever I choose.

As you can see, we have drifted into the subject of the mentally unbalanced. It is not, unfortunately, the one that supplies me with the fewest examples. There are those who make clouds obey their orders, those who can set off earthquakes or volcanic eruptions at will, those who fear the omnipotence of their thoughts.

But let us now turn to a great and serious topic, one that is particularly close to my heart: young people who are strongly drawn to the biological sciences but are kept away from them by lack of mathematical ability. How many saddening letters come to me nearly every day from high school students or their parents, describing a problem that I know all too well!

For example: "I am fascinated by everything about animal life and repelled by everything that has to do with mechanics. My dream is to become a biologist, but I would have to take many examinations that would include mathematics, and I feel incapable of passing them."

Another: "I have loved nature ever since I was a little boy. I dream of flowers, butterflies, beetles, and birds. I would like to know everything about them. Unfortunately, my inadequacy in mathematics practically eliminates me from the courses I would have to take. Isn't there some way I could devote myself completely to the study of animals?"

Still another, from a young girl: "Mathematics is beyond me, and I am interested only in nature, especially insects. Whenever I have any spare time, I observe them. My heart beats faster when I watch ants in an ant hill. Is it really true, as my teacher says, that I have no chance of becoming a naturalist because I know nothing about mathematics and don't want to study it? Are there any schools that teach only biology? I have been advised to study law or literature, but I like only biology and nothing else. What is your advice? If you discourage me too, I will try to accept defeat and stop thinking about insects. But it will be hard."

One more letter, finally, because this is an important matter, and I would be surprised if none of my readers knew an adolescent faced with the same difficulty: "I am seventeen, and I want to study biology, but it seems that my mediocrity in mathematics will prevent me from doing it. I have a distaste for mathematics that I cannot overcome. Without mathematics, is it really impossible for me to devote my life to the study of life? How can I resign myself to giving up the only occupation I would love, the only one in which I would feel at ease?" I could go on quoting letters until tomorrow.

It may be that I am like those medical specialists who imagine that the disease for which they are consulted is more prevalent than it actually is. And, of course, I cannot guarantee that all those young people, incapable of adapting themselves to scholastic requirements, have abilities that would be worth salvaging. Furthermore, I admit that there may be a certain bias in my sympathetic reaction to such letters, in which I find the enthusiasm and openhearted fervor that I myself felt in my youth for insects, tadpoles, and all of nature. Nevertheless, the abundance of concordant testimony makes me more and more firmly convinced that the plight of these adolescents who are denied a chance to fulfill their dreams poses a real educational problem.

Of course, mathematics does help to form the mind, and perhaps it would be better if naturalists were also mathematicians. But all those young people now excluded from biology might very well make up for their mathematical inaptitude by other aptitudes, other qualities, such as adroitness, ingenuity, perseverance, a sense of observation, and something that nothing else can replace: love of the object of their studies.

Defenders of mathematics tend to overlook one fact that shows clearly in those touching letters from adolescents: a naturalist's vocation does not stem only from curiosity and intelligence but also from sensitivity and affectivity. In the words of Konrad Lorenz, one of the greatest naturalists of our time, "There are no good biologists whose vocation was not born of deep joy in the beauties of living nature."

Despite all the objections that were raised against me when I pleaded for "the right to be a naturalist," I remain convinced that our present teaching methods are to blame for creating a large number of unsatisfied, frustrated people who will always regret having been unable to develop their true capabilities. And the biological sciences are deprived of good minds capable of doing fruitful work.

All that discouraged zeal, rebuffed enthusiasm, spurned ardor—are we so rich in scientists, especially those who make significant discoveries, that we can afford such waste? Is it necessary to point out that most fundamental work in biology has been done without any help from mathematics? Cultivation of tissues and organs, transplantation of embryonic cell nuclei, changes of sex, natural and artificial parthenogenesis, the function of hormones, preservation of tissues by cold, the immunological mechanism of resistance to transplants. Not to mention the language of bees and the phenomenon of imprinting in birds.

Unfortunately, there is nothing to justify predicting a better fate for young would-be naturalists in the near future. Mathematical snobbishness (or demagoguery) is now carried further than ever before. Ignorant people who would be mystified by a first-degree equation have decreed that everyone must understand the language of mathematics and know how to use a computer. This is anything but reassuring: since the judgment of even good mathematicians sometimes leaves much to be desired, what can we expect of the mediocre mathematicians being turned out by our schools?

Letters from children make up a significant part of my mail, and they are among the most welcome. Some of them are startling. A nine-year-old boy asks for specific information about parthenogenesis. A little girl asks if it is really true that our ancestors were monkeys. Another, having heard of the seventeenth-century experiment in which Van Helmont thought he had been able to generate mice by placing an old piece of woman's clothing in a container with some grains of wheat, asks if the container must be kept open or closed to make the experiment succeed.

But most of these letters concern little finds that the child has made in nature: snail eggs which arouse his wonder and which he would like to make hatch, a beautiful caterpillar from which he hopes to obtain a butterfly, frog or toad tadpoles that he does not know how to feed properly.

The tadpole plays a predominant part in awakening a child's curiosity. It is often through this creature that he is initiated into the mysteries of animal life. Who has not raised or tried to raise tadpoles in his childhood? Who has not watched the growth of the little webbed feet and followed the entrancing spectacle of metamorphosis? Who has not been disappointed, even saddened, to learn how difficult it is to make the little frogs survive?

How many naturalists' careers have had that simple beginning! When a naturalist relates his childhood memories, he seldom fails to speak of those wriggling commas that animate a pond. The philosopher Gaston Bachelard has said that every school ought to have an aquarium full of tadpoles, because children work better when they are able to see that swarming of life.

I am pleased to see from the letters I receive that love of tadpoles is still as strong as ever among little boys and girls. (According to my personal statistics, even more girls than boys are tadpole fanciers.)

It goes without saying that not all children who are captivated by tadpoles will become biologists, but I believe that by stimulating their innate predilection for animal life we could make many more of them into biologists than we do.

Since I have no secretary, I am unable to answer all my

mail. But I feel it is inexcusable to leave a child's letter unanswered, especially when the awkwardness of its style and the whimsicality of its spelling show that he really wrote it all by himself.

The answer he receives may mean a great deal to him and influence the formation of his mind. It shows him that a grownup has taken his youthful curiosity seriously. That curiosity will then seem more valuable to him, and also to his parents, who may have been inclined to regard it as trivial.

For my part, I never receive a letter from a child without remembering the one that I myself wrote at the age of nine to the great entomologist Jean-Henri Fabre, and recalling the glorious day when the old man's answer arrived.

In speeches, articles, and lectures, I have ardently opposed capital punishment, because I feel, rightly or wrongly, that society should set an example of respect for life by "outlawing death," to quote Albert Camus.

This opinion has brought me many disapproving letters, often of a particularly violent kind: "So you have no pity for murder victims or their relatives! All the compassion you are capable of is reserved for monsters." I am all but accused of approving the most hideous crimes. Some go so far as to write, "It's not surprising that you're on the side of murderers, since you spend all your time murdering frogs." Or: "You would do better to feel sorry for your frogs, rather than for murderers." Always frogs.

The reproach sometimes even takes the form of a homicidal wish: "I hope that some day you or someone dear to you, a member of your family, will be the victim of one of those people you support. Maybe then you will change your mind." Or, still more explicitly: "I hope that you will encounter one of those vicious murderers who have been spared by the excessive leniency of our courts, and that he will show you no mercy."

I will refrain from concluding that supporters of the death sentence are more ferocious than its adversaries. Malice is not limited to either side, and written words are sometimes harsher than the writer's real thoughts.

The matter of the extra chromosome—that famous chromosome, which, when it is doubled, creates a predisposition toward criminal behavior—naturally appears in the mail of a biologist who opposes the death penalty.

"Another trick by scientists to save the necks of criminals." . . . "We don't believe one word of that nonsense about chromosomes." . . . "If we start examining everything— blood, saliva, urine, cells—we will always find something wrong, and then no one will ever be responsible for anything." . . . "Are you sure you don't have a chromosome missing yourself, the chromosome of common sense?"

Ah, how hard it is to please everyone in these moral questions! Having said in a lecture that science ought to prolong human life as much as possible, and then that measures should be taken to prevent population growth, I was vehemently accused of wanting to sacrifice babies for old people. Having spoken of the danger that a society would face if it legalized the elimination of deformed babies and incurable invalids, I received indignant protests, some of which were disturbing because they came from invalids, their relatives, or their nurses: "If you knew the suffering that can be caused by stubborn efforts to prolong lives that are nothing but pain, you would not speak as you do."

Some of my readers may recall the deplorable affair of Naessens, the charlatan who claimed to have invented a miraculous cure for leukemia. When I had taken an opportunity to say on radio what I thought of his claims and methods, as well as of certain publications that gave false hopes to anguished relatives of leukemia victims by supporting him, I received only letters full of protest and abuse.

"You, too! You condemn a great scientist because he works outside the official framework! I thought you had a more independent mind. Your prejudice is appalling. It must be the French Academy that has made you so hidebound. France will soon regret the way she has treated Naessens. He will take his genius to another country, and his glory will still be shining all over the world when your frogs have long since been forgotten."

Needless to say, when Naessens was convicted of fraud, no one wrote to acknowledge that I had been right. Many, in fact, wrote in this vein: "Even if Naessens was a charlatan, you should not have said so, because he was helpful anyway. He gave hope to people who had lost it."

Thus one is regarded as cruel for having denounced a lie.

Moral truth is often difficult to determine (Jankélévitch has spoken of "the moral indeterminate"), and even scientific truth gives rise to debates that may be distressing to anyone not accustomed to seeing it falsified for motives alien to science.

Beginning about 1948, there was much talk in scientific circles about the extraordinary assertions of the Soviet botanist, Lysenko, and his disciples, who had set out to destroy orthodox genetics (or "Mendelo-Morganism," to use their term) and replace it with a new "Michurinian" genetics, of Marxist and proletarian inspiration.

In opposition to "bourgeois" geneticists, they maintained that acquired traits were transmitted to offspring, that chromosomes were only a figment of the imagination, that wheat could engender rye or barley, that weeds were descendants of useful plants, that cells appeared in the yoke of an egg by spontaneous generation.

This rehashing of medieval concepts, presented as the latest word in biological science, was ardently supported by many intellectuals of the far left. A good Communist had to believe in Michurinianism, which in turn, by its discoveries, demonstrated the fecundity of Communism.

It was easy for any unbiased biologist to see that this was only a case of collective delusion. As usual, I stated my opinion. I said how sorry I was to see political prejudice being introduced into what should have been a purely scientific controversy. Astonished, disappointed, and accusing letters immediately began pouring in.

"We thought you were in favor of progressive ideas, but now you have sided with conservative, bourgeois science. You are playing into the hands of our adversaries. I did not think your mind could be so clouded by class prejudice."

I admit that I was hurt by these letters, because I knew I did not deserve the reproaches they contained. And they were all the more painful to me because in my innermost heart I could fully understand the feelings that prompted them.

Those who condemned and insulted me were no doubt sincere people, friends of truth who, if they had been in a position to determine the facts, would have agreed with what I said. But since they had no means of forming a correct opinion, they could only follow those scientists who vouched for the truth of the "Michurinian" doctrine in the name of Marxist dialectics.

If I had been in their place, I would have been misled and deceived as they were; I would have reacted as they did; I too would have thought that a revolutionary biology was being rejected because of ideological resistance.

I have spoken of the atomic danger on many occasions. I have described the risks imposed on the human race by the criminal dissemination of nuclear weapons. I have said that every nuclear explosion, no matter what precautions are taken, raises the level of ambient radiation and, as an inevitable consequence, increases the rate of birth defects and abnormalities, as well as of leukemia and cancer.

It seems perfectly natural to me that my statements and acts should arouse criticism and objections. No one is more willing than I to have my protests protested. But what surprises me and seems completely wrong to me is that some of my critics try to deny me the right to express my opinions on nuclear armament with animation, fervor, and passion.

"You speak as a partisan, and that is unworthy of a scientist. From now on, I will be suspicious of everything you say. Even when you speak about biology, I will no longer have any confidence in you. Even more than what you say, it is your tone that disappoints and shocks me. You could express the same ideas calmly, sensibly, and dispassionately, without stooping to demagoguery."

I confess that I do not see why a scientist—who is also simply a man, a citizen—should not have as much right as anyone else to engage in a struggle that seems right to him, and commit himself to it entirely, with his sensitivity, his temperament, his feelings—his whole being, in short—especially when he is concerned with great moral issues that go beyond politics in the narrow sense and ultimately involve choices that are not purely rational but are also emotional. I cannot help it if I do not have a lukewarm temperament.

Among the reproaches addressed to me because of my political opinions, there is another that I reject: "It is unfair to aid a political cause with a reputation earned in another field." It is hard for me to understand why a modest reputation as a biologist should be regarded as a muzzle.

Still another objection that I consider unacceptable: "Is it proper to have such opinions when you are the son of Edmond Rostand?" First of all, I do not know—and neither does anyone else—what my father would think today; his work is rich enough to provide a number of different ways of remaining faithful to it. And then, no matter how great my filial piety may be, it could never hinder me in exercising an independence and frankness that I learned from reading *Cyrano de Bergerac*.

Finally, there are more amusing reproaches. Some come from people who, thinking they are being cruel, direct their remarks against my age; past seventy, it is not good to hold a "subversive" opinion. They jeer at my senile fervor, they claim to be saddened that I should degrade the dignity of my white hair by flying into unseemly rages. A respectable psychiatrist advises me to withdraw from active life and gives me the name of a rest home for tired intellectuals.

And a self-styled professor of history writes to me: "In your criticism of the nuclear deterrent force, you claim that an increase in atmospheric radioactivity can produce a whole generation of monsters. As far as I know, there was not yet any atomic pollution at the time when you were born. But take a look at yourself in a mirror." How pleased the "professor" must have been with himself when he wrote that.

Let us leave politics for occultism, which I have attacked all my life, in all its aspects and forms, from parapsychology to astrology, passing by way of belief in the power of divining rods.

As you can well imagine, sorcerers and magicians have not remained silent about me. Half a century ago there was much discussion of psychic mediums in certain intellectual circles.

Great scientists—William Crookes, Charles Richet—stated that they had ascertained the existence, in certain exceptional individuals, of supranormal powers manifested by phenomena that violated all the laws of physics: the ability to move objects at a distance, mysterious knockings, luminous apparitions, emission of a viscous substance, or ectoplasm, that took the shape of a foot, a hand, or a face.

That kind of occultism has now died down and ceased to be fashionable, which is certainly a substantial gain for friends of reason, but it would be a great mistake to assume that occultist folly has laid down its arms. Although interest in the physical feats of mediums has declined, and we no longer hear much talk about levitation, phantoms, or ectoplasm, public opinion is more favorable than ever toward parapsychological phenomena—telepathy, premonitions, clairvoyance—and the false sciences of divining and astrology. "Magnetic healers" and fortune tellers do a thriving business; newspapers are obliged to give their readers the shameful fare of horoscopes to avoid a drop in circulation. This neo-occultism is abundantly represented in my mail.

First, there are those who, quite courteously and sincerely, offer to give me the benefit of their knowledge or volunteer to serve as experimental subjects. They are so confident of having supranormal faculties that they accept all experimental controls in advance and guarantee to convince me if I will only consent to verify what they say.

A young woman tells me that she can detect all ailments by swinging a pendulum that she has made for herself. (She has swung it over a newspaper photograph of me and can give me some valuable information about the state of my heart and liver.) A clairvoyant man says that if I will show him the backs of playing cards he will guess them correctly three out of four times. (He will bring his own deck of cards, but it can be examined by a prestidigitator.) Another man can soothe

the most stubborn pains, no matter what their origin, by applying to the painful area a wad of cotton that he has previously "magnetized" with his left hand. (If he magnetized it with his right hand it would increase the pain rather than relieving it.) Another, by means of his "fluid," can accelerate the growth of a potted hyacinth. (It must be treated every day, preferably in the morning.) Another proposes to send me a telepathic message once a week, between six and seven o'clock in the evening, and asks me to acknowledge receipt of it. Another—a policeman, and full of solicitude—knowing that I was born under the sign of Scorpio, urges me to beware of the quartile of Saturn for the next few months.

But there are also less friendly letters. "How dare you make a blanket condemnation of the esoteric sciences when you do not know the first thing about them? Rejecting all of occultism is unworthy of a mind that claims to be free and scientific. It shows a dogmatic, sectarian narrow-mindedness that cheapens the value of your judgment in other matters. You are nothing but a false scientist, a sham biologist. What would you think of an ignorant layman who made categorical statements on genetics or embryology and had dogmatic opinions about what happens with your frogs? That is exactly what you do when you pronounce judgment on divining and astrology."

My skepticism is interpreted as a personality disorder: "You are a born denier, a constitutional doubter. You would go on quibbling before the most solidly demonstrated fact in the world. It is not your fault that you have such a mental problem, but you ought to see a psychiatrist about it." Or, still more acridly: "Your limited intellect makes it impossible for you to understand anything immaterial or imponderable. In spite of your learning, you are only a conceited ignoramus, an incorrigible materialist, a crude rationalist, a biologist barely qualified to be a laboratory assistant. You spout scientific platitudes and reject all mystery to protect your grubby philosophy.

"And do you know that your scientistic intransigence is an outmoded, obsolete attitude? Read *Planète** and you will begin to educate yourself. You will learn that a doctor at the University of Pennsylvania has demonstrated the reality of telepathy and that even the Russians use it to guide their submarines."

Inevitably, an allusion to Edmond Rostand: "You deny the marvelous because you did not inherit the gift of poetry from your father. Without it, no one can reach the highest realms of thought. Your father enchanted us with his eagles and cocks, which, you must admit, had much more style than your toads and frogs."

In all this drivel, one delectable letter: it is from a man who assures me that in amorous relations the transmission of thought is so obvious that anyone who denies it reveals his emotional poverty. "It is nothing for you to be ashamed of, but allow me to pity you."

And finally, a letter from a young woman whose gentle melancholy, I must confess, saddened me a little: "I have just read your article on astrology. It is possible that you and Paul Couderc are right in what you say about horoscopes. But why do you want to take away my illusions?

"To live, I do not need truths, but I do need illusions.

"My horoscope announced a favorable period in April. I counted the days and waited anxiously. Now, because of you, I am like someone who has lost his reason for existing.

"Scientists may be fine men, but I prefer poets, and that includes astrologers."

Truth versus illusion; we have already seen that opposition

* TRANSLATOR'S NOTE: A French periodical that publishes articles on esoteric subjects.

in the Naessens affair. Always the same problem: Does a comforting illusion deserve to be treated with respect?

I have committed a more serious crime than poking fun at magicians: I have contested the philosophy of Teilhard de Chardin. Not that I do not admire the author of *The Phenomenon of Man.* I respect him as a paleontologist, as a moralist, and as a profound and lyrical writer; but I once ventured to say that his concept of evolution brought us nothing new and that there was a certain exaggeration in the importance given to his thought.

It was as if I had stepped on a hornets' nest. Teilhardians— and especially, perhaps, the women among them—are irascible people. "You are not worthy of speaking about Teilhard. You are jealous of his influence on young people. Of course you cannot understand him, since your mind is forever closed to spiritual realities, and it is lucky for him that you cannot. But you committed a bad act when you attacked that great consoler. And you caused me great pain. I will never again open one of your books. I no longer admire, like, or believe you."

And that brings me to my final point, the gravest, most serious of all. I will surprise no one when I say that I have often been reproached for what is called my disheartening materialism. I might argue about the word "materialism," because those old philosophical classifications no longer have much meaning, but I readily admit that my opinions on the meaning and scope of the human adventure are not particularly bracing and comforting. I therefore see nothing startling in the fact that they sometimes arouse opposition and disapproval.

"You reduce us to the condition of animals. To you, we are nothing but bits of dust about to return to the earth. Your nihilism is an offense against human dignity, an invitation to despair. It is truly shameful to flaunt such opinions. If you really believe those depressing things, at least be considerate enough to keep them to yourself. You should not sow confusion in people's souls, you should not make life harder for those who are suffering. Think whatever you like, or whatever you can (it it not your fault if you are made that way), but if that is all you have to tell us, it would be better to say nothing. Don't you love anyone? If you do, how can you accept the idea of permanent separation? Do you consider it one of the benefits of science when you make human distress still greater by negating all hope?" . . . "Do you feel that you are doing something worthwhile in propagating your philosophical nihilism? For some reason you are apparently able to bear it yourself, but it can be harmful to other minds. Please keep your poisons for yourself." . . . "You have often opposed a certain kind of demoralizing literature, but what could be more demoralizing than your thoughts? Even abjection is better than despair, which is perhaps all the more dangerous when it is nobly accepted." . . . "You protest against the atomic bomb and the madness of war. But what is the good of stubbornly defending human life if it has no meaning and ends with the grave? What is the good of scientific effort, of unrelenting pursuit of new truths? In such a bleak context, your frogs cut a very sorry figure, and I hope you realize it." . . . "If you really believe what you say, if your vision of man is really so dark and limited, how do you go on living, working, continuing your research? At what hearth do you warm yourself? What gives you the strength to carry on? In short, from what 'shadow of a shadow' do you live?"

These are obviously harsh questions, and an attempt to answer them, if only for myself, might take up all the little time I have left to live. They are so harsh, and so serious, that it might have been better if I had stopped a little sooner in my inventory of a biologist's mail.

Such letters never fail to disturb and distress me; they rekindle old misgivings in me; they repeat what I sometimes tell myself. One of them, in particular, placed me under a great strain: it was from a young man who begged me to tell him if I was really convinced that there was nothing after death, and it said that the whole course he would give to his life would depend on my answer. Those few lines turned a vague, anonymous reader into a flesh-and-blood human being, who was holding me responsible, so to speak, for the use he was going to make of my thoughts. I would have preferred, of course, not to answer.

It is painful to be a destroyer of hope; it is not pleasant to be regarded as a man who sows despair. The idea of making life harder for someone is not easy for me to bear. But I also feel that any sincere, disinterested conviction has a right to be expressed. I believe, in fact, that it creates a duty to express it. Is it not important for those who think as we do to know that they have brothers?

I hope that in these few pages I have given a fairly accurate idea of what a biologist's correspondence can be like. I believe I have shown that it is varied, curious, strange, and sometimes disturbing. It contains many reproaches—and that is not surprising, considering the number of things I have attacked or disputed in the course of my long life.

I have no doubt given a little too much space to letters of criticism and even abuse, not out of some sort of masochism, but because they make livelier reading than praise. Now and then, however, I receive a friendly letter that comforts and reassures me. Someone tells me that one of my books inspired him with a vocation that has become the center of his life, or showed him a similarity of thought that was beneficial to him.

This encourages and compensates me; it gives me new zest for continuing on my way. When it comes at the right time, a letter from a student, an old schoolteacher, or a young nurse may tell me just what I needed to hear. But such things are best kept to oneself.

PRESENT AND FUTURE OF THE HUMAN PERSON

The word "person" comes from the Latin *persona*, a word of unknown origin, which, according to the dictionary, "first designated the mask worn by an actor, then, by metonomy, the role of an actor, the character portrayed by him. Finally it came to signify, in general, the idea of individuality, of personality."

As we know, this idea of personality, of individuality, plays an important part in medicine, psychology, education, criminology, morality, philosophy, politics, literature—and above all in the daily experience of life, in which it dominates interhuman relations.

It is this ordinary experience that I now want to use as my point of departure, deliberately leaving aside the ancient debates of the realists and nominalists concerning the principle of individuation, egoity, ipseity, and other scholastic intricacies.

What, then, is a "person," in the usual sense of the term?

It is an extraordinary composite of body and mind, a psychosomatic mixture. It is a face, expressions, a smile, a look, a tone of voice, familiar gestures, a walk, a handwriting (that "living portrait," as Marceline Desbordes-Valmore said), a sensitivity, a character, a disposition, a turn of mind, a past. It is a whole world—in short, an inexhaustible microcosm. Who would dare, even with the meticulous talent of a Marcel Proust, to undertake a complete inventory of a person, even the simplest, most ordinary, commonplace, transparent, and easily fathomable person in the world?

By way of a preamble to this subject, I would like to quote a passage that has always seemed to me extremely striking and moving in its starkness, a passage, written in 1796, in which the great mystic writer, Novalis, sketched a description of a young girl: his fourteen-year-old cousin, Clarissa, whom he regarded as his fiancée. She died a year later. Few other pieces of writing give us such a vivid feeling of entering into the miniature universe constituted by a human being.

"Her precocious maturity. Her attitude toward sickness. Her whims. What does she like to talk about? Her judgments. Her opinions. The way she dresses. Dancing. Her activity in the house. Musical ear. Her taste. Her features. Her face; her vitality; her health; her political position. Her movements. Her language. Her hands. What does she like to eat? Her way of being glad or sad. What pleases her most in a human being, in an object. The tobacco she smokes. Her fear of ghosts. Her thriftiness. Her face when she hears an off-color remark. Her talent for imitation. Her generosity. She is irritable, touchy. Her dislike of teasing. Her attention to the judgments of others. Her spirit of observation. She is terribly afraid of mice and spiders. She does not let others address her with familiarity. A birthmark on her cheek. Her favorite foods: herb soup, beef, beans, eel. She readily drinks wine. Likes the theater, comedies. She meditates much more on others than on herself."

Yes, I admit that this passage seems to me extraordinarily
evocative, because it shows in rudimentary form what a complete, exhaustive portrait of a person might be. In that bare
enumeration, with everything placed on the same level—the
mental and the physical, the important and the incidental,
the profound and the superficial—in that brief summary which
Novalis could have extended to ten thousand pages without
exhausting his subject, I see an illustration of Leibnitz's
thought: "Individuality contains infinity in the bud, so to
speak."

Is it necessary to add that such lines could have been
written only by someone in love? For only a lover can give so
much attention and value to another person's slightest traits
and peculiarities. We make no choices in what we love, we
take everything together. Love is the surest and most sensitive
reactive agent for individuality. We do not, however, love
everything in the person we love, and this is the source of
most misunderstandings and tragedies in love.

How many quotations come to mind with regard to human
personality! The singularity of the person could be the subject
of an excellent anthology.

MONTAIGNE: "Because it was he, because it was I . . ."

PASCAL: "Diversity is as wide as all tones of voice, all walks,
coughs, nose-blowings, sneezes."

VIGNY: "Love what you will never see twice."

George Bernard Shaw belittles individual singularity with
his remark that to love a woman means to overestimate the
difference between one woman and another, while William
James exalts it when he says that there is little difference
between one person and another, but that that difference is
everything.

The diversity of human faces alone has always stirred the
curiosity of thinkers and inspired writers.

PLINY: "Although, in man, the face is composed of no more than ten parts, there are no two identical faces among all the thousands that exist. The art of nature, despite all its efforts, cannot achieve that diversity in the greatly limited number of its combinations."

Bernard de Fontanelle wondered, "What secret nature must have, to vary something so simple as a face in so many ways."

The anatomist Nicolas Leméry wondered at how different faces were from one another when all were formed on the same model, that is, on the same number, nature, shape, and arrangement of parts. The differences are such that if, within the whole multitude of the world's population, two faces should ever be found which resembled each other so exactly that they could be examined side by side without revealing any way to distinguish between them, it would be one of the rarest and strangest natural phenomena ever seen.

According to Leméry, the variety of human faces is part of nature's plan; it is willed by the Creator as a necessity of the social order. For if all men were "so perfectly alike that nothing distinctive could be seen in any of them, how would they recognize each other? They would pass before each other's eyes without seeing, or at least distinguishing, each other; they would be as incapable of it as if they were blind. They would always be losing one another without being able to regain contact, and these continual tribulations would make them hate society all the more, because it would then lack the benefits it provides for them now, as things actually are."

The discerning Joseph Joubert contents himself with remarking, "It is chiefly by his face that a man is himself."

One of our illustrious contemporaries, François Mauriac—in *Ce que je crois*, a work in which he reveals the depths of his thought to us—has evoked the astonishment and wonder aroused in him by the spectacle of the diversity of faces:

"There is a miracle so common that we no longer even see it: of all the human faces that exist and have existed, none is the same as another. There are no two identical faces in nature. Not one face duplicates, feature for feature, that of any of the billions of people who have lived before us. Each human being is a unique specimen that has never been duplicated since the world began. This singular, irreplaceable nature of even the humblest human creature is an obvious fact. It prevents us from confusing people with one another, it enables us to recognize them in a crowd . . . and it helps me to understand how each of us can be the hero of that drama of salvation in which eternity is at stake."

The biologist Vandel draws a moral lesson from the singularity of each human being: "A man is not one of many interchangeable representatives of a species, but a person unlike any other, and consequently irreplaceable. To eliminate one man is, to a certain extent, in a certain way, to impoverish mankind."

And the philosopher Schopenhauer wrote, "The deep grief that we experience at the death of a friend comes from the feeling that in each individual there is something indefinable, characteristic of him alone, and therefore absolutely irreplaceable. *Omne individuum irreparabile.*"

We find that same feeling in an admirable passage in which an eminent French surgeon, Professor Hamburger, has described the reflections induced in him by a little girl named Nicole, on whom he was about to perform the hazardous operation of a kidney transplant: "I remember that sickly little girl, her apprehensive look, her pale complexion, her features so deeply marked by suffering. Should we have resigned ourselves to letting her life end, on the grounds that her nine brothers and sisters were more than sufficient to perpetuate the family? From the deepest roots of our vocation as doctors, we feel that it is impossible to consent to such an attitude. Our rule is simple and straightforward: to preserve life at all cost, and not the life of the species, but the life of the individual. It is true that little Nicole is nothing, nothing but a defective link, nothing, in all likelihood, of any pragmatic interest to the species. But she is nevertheless irreplaceable. I do not know exactly why she is so valuable, why I am so strongly affected by the idea of her death, when I know that it is inevitable sooner or later, and why each drop of that life is so precious, why each hour gained is so necessary. Perhaps little Nicole is irreplaceable for the sole reason that she is unlike anyone else. No other little girl, not even her twin sister, has exactly the same mind, thoughts, sensitivity, and inner world as Nicole. That is why the foundation of our ethic is simple. The judge may complain that justice is difficult to define, the politician may be hesitant about the principles of his action, the archaeologist may choose among a score of different projects; our own struggle has only one goal: the health and life of man taken as an individual, as a unique individual. We have no need to philosophize on the meaning of that life, on its value to the community, on its place in human continuity. To us, even the most fragile, precarious, and useless life is of infinite value." (*Bruxelles Médical,* October 8, 1961.)

The biologist Cyril Dean Darlington says that recognition of human individuality ought to be the basis of all legislation. As for the sociologist Jean Fourastié, he hopes that when the society of the future establishes its rules, it will take account of that originality of each person which requires a certain individualization in collective solutions: "Objective recognition of the diversity of men's economic, emotional, philosophical, esthetic, and spiritual needs must lead the society of the twenty-first century to tolerance and a concomitant diversity of products, human relations, activities, and interests."

We have just seen individual singularity—the singularity of the person—as an object of curiosity, surprise, wonder, love, concern, solicitude, and respect. For the believer, it supports religious conviction; for the biologist and the physician, it reinforces respect for life; for the sociologist, it stresses the need for tolerance and justifies the wish to secure the individual against the despotism of the group; for the philosopher, as for everyone else, it accentuates dismay with regard to death, which abruptly draws a line that crosses out the infinite.

It is now time to ask where that individuality comes from and how it is formed. What is it that makes a person himself? A few biological details are indispensable here, because individuality begins at the moment of conception.

Every human being comes from an initial cell, the fertilized ovum, which is formed by the conjunction of two cells, one from each parent: a female germ cell, or ovum, and a male germ cell, or spermatozoon.

In each of these cells is a nucleus enclosed in a membrane, containing a fixed number of minute particles, the chromosomes. There are twenty-three of them in each cell; the fertilized ovum therefore contains forty-six, in twenty-three pairs, each composed of one paternal and one maternal element.

It has now been established beyond doubt that the chromosomes are the principal agent of heredity and are thus largely responsible for making each person what he is. They have this power because they contain a very great number of molecules of a highly complex acid, deoxyribonucleic acid, or DNA, about which we have recently heard much talk. We are beginning to understand the structure of DNA. Modern biochemistry will have achieved one of its most astonishing successes when it has thus clarified the nature of the material elements that contribute so greatly to making each of us himself.

Each DNA molecule consists of two very long strands twisted into a double helix and composed of a chain of elementary units (nucleotides) characterized by the presence of the following organic compounds: *adenine, guanine, thymine,* and *cytosine.* Adenine and guanine are purine bases; cytosine and thymine are pyrimidine bases.

The terminology may be a little forbidding at first, but the names of those four bases must be repeated again and again, to make everyone familiar with them. *Adenine, guanine, thymine, cytosine:* these words must become part of everyday language, as is already the case with "gene" and "chromosome." No one should decline to learn that elementary bit of chemistry, because it is essential to any further understanding of man.

Oscar Wilde said that heredity is the only god whose name we know. We can now do better than know its name; we know its formula! This means that we can now extend into the invisible our analysis of that "certain something," that "momentous trifle," which, as Pascal said, can have such frightening effects when it produces love.

If a purine base had been displaced in a molecule, Cleopatra's nose would have been shorter, and the whole face of the world would have been different!

The properties of an individual's heredity depend on the way in which those four bases are arranged and ordered in their molecules; all the genetic diversity of the species stems from them, just as all our literature is written with twenty-six letters, and all our music with seven notes.

Since there are several million of these bases in each fertilized human ovum, the number of their possible arrangements is so great that it is practically impossible for genetic combinations, formed by chance, to result in two ova with the same molecular structure.

Thus we can say that every fertilized human ovum has a chemical endowment, that is, a chromosomal makeup, that belongs to it alone. Each person begins his existence as the only one of his type—he is *unique*. He will play the game of life with a hand that has never been dealt to anyone else. And even if the human race lasts for trillions of years, it will never make any genetic repetitions; no two individuals with the same heredity will ever appear on the planet.

This is one of the great facts of biology, and it can never be overemphasized. A human being who exists only in the form of a microscopic cell is already distinct from all others, already unique; the material basis of the individual self is already firmly established.

In the collection of molecules that he has inherited from his parents, a large part of each person is irrevocably inscribed and determined in advance: the shape of the features of the face (that face whose diversity has intrigued so many thinkers); the color and quality of the hair; the shape, length, and density of the eyelids and eyebrows; the skin color; the pattern and color of the irises; the size, shape, and folds of the tongue; the size and shape of the ears; the shape and alignment of the teeth; the pattern of the lines of the hand and the dermal papillae of the fingers; the blood type; the Rh factor, etc.

Genetic uniqueness is notably manifested by certain physical traits, such as fingerprint patterns, which begin to appear in the fourth month of fetal life. Anyone who has seen a French identity card knows that there is a little rectangle on it for a fingerprint, usually of the left forefinger. This method of identification is based on the fact that no two individuals ever have completely identical fingerprints.

Dr. Victor Balthazard, an expert in forensic medicine, has pointed out that an average of one hundred particularities can be distinguished in every fingerprint. To have a reasonable chance of finding two prints that match in sixteen ways, it has been calculated that one would have to examine 4,294,967,296, a number greater than that of the earth's inhabitants. Beginning with seventeen coinciding features, the chances become practically nonexistent; in other words, it can be assumed that any two prints that match in at least seventeen ways were made by the same person.

Moreover, fingerprints are not the only means of identification: one can also use an imprint of the palm of the hand or the sole of the foot, or a carefully selected set of structural characteristics.

Alphonse Bertillon is often credited with being the first to have the idea of using fingerprints. Actually, the method was first promoted by Faulds in Japan (1878) and Francis Galton in England (1888), and Bertillon gave it his support only toward the end of his life.

While a person's physical characteristics depend to a large extent on the chemical endowment constituted by the germinal nucleic acids, it is obvious that they also depend greatly on the way he lives and the effect of circumstances. His size, for example, is partly determined by the amount of food he eats in early childhood; his muscles are developed by exercise, etc.

With regard to his moral and intellectual traits, the part played by external factors is much greater. Although his individuality is partially formed by the nucleic acids even in this respect, it is easy to understand how his mind, sensitivity, and character can be influenced by upbringing, culture, education, social milieu, family climate, emotional relations with parents and brothers or sisters, friendships, acquaintances, films and plays, reading, and so on.

We must not overlook the physical and emotional state of the mother during pregnancy, the first sensations of the newborn baby, the first faces he sees, the way he is fed and weaned, and even the name he is given.

Concerning this possible influence of a name, I will quote a curious and little-known passage by Bernardin de Saint-Pierre.

"A child," wrote the author of *Paul and Virginia*, "patterns himself on his name. I have known unfortunate children who were teased so much by their playmates, and even their relatives, about their given names, which had certain connotations of simplicity and good nature, that they gradually took on the opposite characteristics of shrewdness and ferocity."

Without subscribing to that interpretation, I am willing to grant that *anything* may act on an individual—anything except the position of the heavenly bodies at the time of his birth!

Let me point out, furthermore, that there is a constant interaction between the physical person and the moral person. Temperament and character depend partly on such general feelings as vigor or lethargy, and even to some extent on what one sees in the mirror. There will be personality differences between a tall, robust man and a short, sickly one, between a very ugly woman and a very beautiful one.

On the other hand, character has a certain influence on physical appearance. It has been said that, past a certain age, everyone has the face he deserves. This is no doubt an exaggeration, but a person's exterior is shaped and animated by his interior. Foolishness, malice, bitterness, pettiness, and ill humor are inscribed in the face, and so are their opposites. But it would be an endless task to list all the factors that may cooperate with heredity in fashioning the individual.

Briefly, each of us is what he is because he started from a certain ovum and because he has led a certain life. He is doubly unique: by the singularity of his origin and by the singularity of his personal adventure.

Let us recall Novalis's description of his young fiancée; it is probable that her "musical ear" was inscribed in her genes, but for everything else, who could disentangle what was caused by her nucleic acids and what was the effect of circumstances?

I have emphasized the part played by the chemical personality of the original cell in the genesis of the person. This chemical personality is maintained through all the cellular divisions that take place in the organism, beginning with the ovum, and is thus found in each of the billions of cells that compose the individual. Peter's blood corpuscles, the cells of his epidermis and glands, the fibers of his muscles, and the neurons of his brain differ, by their nucleic acids, from Paul's blood corpuscles, epidermal and glandular cells, muscle fibers, and neurons. Peter and Paul are themselves—and unique—in even the least of their elements.

Furthermore, this identity will be preserved all through their lives, despite the renewal of tissues (some of which are replaced very often), senile decline, changes of appearance, illnesses, accidents, medical treatments, and even blood transfusions. From conception to death, the biological personality is constant, invariable; we all remain faithful to ourselves to the end.

In some people, however, those known as "mosaic individuals," the body contains parts that are not in conformity with the rest of the person and do not correspond to the genetic determination of the original cell. This is because, in the course of their development, a change has occurred in the chromosome composition of one of their cells (somatic mutation), and all the descendants of the mutant cell have inherited the mutation. Thus a person may have eyes of two different colors because of a mutation affecting the cells that formed one of the irises.

Accidents of this kind may affect the chromosomes that determine sex, producing sexually heterogenous individuals who have a mixture of male and female tissues; they are comparable to those strange butterflies called gynandromorphs, which have male wings on one side and female wings on the other.

Other mosaics combine normal tissues with "mongoloid" tissues. Some have been discovered which combine three or even four types of cell populations; and we know only those that can be easily detected by examination of the chromosomes—how many others, more elusive, must pass unnoticed! The discovery of these individuals who are genetically "several in one" is one of the important new achievements of human biology.

It seems rather likely that malignant tumors, or at least some of them, are caused, like mosaics, by somatic mutations, except that the mutations take place at a later stage of life. If so, the newly formed cellular minority is endowed with aggressive properties and has the fatal power to destroy the majority of the individual's cells.

As we have just seen, there are people who are "several in one"; there are also some who are "one in two"—identical twins. The human species produces two kinds of twins, or individuals born of a single delivery. Fraternal twins come from two different ova, fertilized by different spermatozoa. Identical twins come from the same ovum, fertilized by a single spermatozoon, which splits in two at a certain stage of its development.

Fraternal twins are about two and half times as numerous as identical ones; that is, fraternal twins are born in about one out of every eighty deliveries, identical twins in about one out of every two hundred.

A human ovum sometimes produces more than two identical offspring: three, four, and even five, as in the famous case of the Dionne quintuplets in Canada.

Fraternal twins, of course, have dissimilar genetic makeups; each twin has his own distinct biological personality and uniqueness. They are basically two ordinary brothers or sisters, or an ordinary brother and sister, since they may be of different sexes. One may be dark-haired and the other blonde, one may be tall and the other short. Identical twins, however, are always of the same sex and resemble each other amazingly, even in the slightest details of their morphology and physiology. They are "the same individual published in two copies," to use Dr. Apert's apt expression.

Pascal was surely thinking of identical twins when he wrote, "Two like faces, neither of which makes us laugh when we see it alone, make us laugh when we see them together, because of their likeness." Bergson comments on this remark in the light of his theory of laughter: "Truly living life should never repeat itself. Analyze your impressions when you see two faces that resemble each other too closely: you will find that you think of two castings from the same mold, or two imprints from the same stamp, or two photographs from the same negative; in short, some sort of industrial process. This deviation of life in the direction of mechanics is here the real cause of laughter." (*Le Rire.*)

There is the case of two orchestra conductors who were identical twins; one could take the place of the other in the course of a concert without anyone in the audience being able to detect the difference.

Even in fingerprints—a preeminently individual characteristic—the resemblance between identical twins is generally quite pronounced. Nevertheless, fingerprints can be used to distinguish between identical twins whose likeness might create difficulties. Sannié tells of a woman in Indiana whose twin daughters looked so much alike that she was afraid of being unable to tell them apart. She had them both fingerprinted by the Evansville police, and from then on there was no longer any risk of confusion.

I am reminded here of Mark Twain, who said he was not sure if he was still alive, because while he and his twin brother were babies, their mother had got them mixed up in their bath, and the brother had later died.

It goes without saying that the resemblance between identical twins is no reason for assuming the existence of some sort of mysterious psychic communication between them; I hope no one believed what was recently stated in an evening newspaper: that when one twin sister cut her finger, the other, far away, felt pain.

There is, of course, great biological, physiological, and even philosophical interest in the study of identical twins, since they both have the same origin and began as a single individual. In some cases it enables us to distinguish between hereditary and environmental factors in the formation of the person. One twin is obviously the perfect "control" in relation to the other.

We must bear in mind that even when identical twins grow up apparently under the same circumstances, this is never completely true; they did not both occupy the same place in the uterus, one has had an illness that the other has not had, one has read a book that the other has not. Their origin is the same, but their lives are always different, to some extent.

Precisely because the existence of identical twins breaks and seems to challenge the great law of the biological uniqueness of the person, it accentuates that uniqueness and calls attention to it. The fact that two twins share the same biological self reminds the rest of us that we alone possess ours, that each of us is the only one of his kind. And if the theme of twins has been so abundantly exploited by writers, especially playwrights—from the ancient Greeks, such as Antiphanes, Anaxandrides, Aristophanes, Xenarchos, Alexis, and Menander, to Jean Cocteau, Jean Giraudoux, Sacha Guitry, and Jean Anouilh, passing by way of Plautus's *Menaechmi*—it is not only because it is a source of amusing mixups, but also because it concretizes the absorbing notion of the biological personality.

"If, in our fables and legends," says the psychologist René Zazzo, "we so often play with the resemblance of twins, and if we so often make it an object of ridicule, it is no doubt in order to rid ourselves of the uneasiness it arouses in us."

And further: "Each man's attitude toward the idea of a double, a twin, is much more complex than a reaction of intolerance. It is made up not only of anxiety, but also of desire, revolt, and a strange fascination. This is no doubt because the question 'to be or not to be' presents itself to everyone, even the least metaphysical of men. The idea of a double is, however, an ambiguous answer to that question. It contains a threat of alienation and disintegration, as well as the promise of a discovery, a seizure of oneself."

Concerning the emotional relations between twins, Zazzo has called attention to personality disorders related to their situation. In general, twins are bound together by "an astonishing love," but sometimes there are reactions of aggressiveness, even rebellion, against the too similar companion. The presence of a "double" exacerbates narcissism and makes construction of the self more difficult. There is a conflict between "the pleasure of resemblance and the need to be a person."

In her *Memoirs of a Dutiful Daughter*, Simone de Beauvoir tells us that it would have been hard for her to bear the existence of a twin sister who would have deprived her physical self of "what gave it all its value: its glorious singularity." Existentialism is not very willing to share.

Until recent years, the organic sameness of identical twins was a dogma of biology. We now know that there are very rare exceptions to that rule.

When the ovum divides to produce two identical twins, a

mutation may occur in one of the halves. Thus Dr. Jérôme Lejeune has reported the case of a pair of identical twins, one of whom was a sexually normal male, while the other exhibited female characteristics; the cell from which the latter developed had lost the Y chromosome, which determines masculinity. This was an instance of the same type of accident that we have already mentioned in connection with the formation of mosaic individuals. If genetic plurality is possible within a single individual, why should we be surprised to find it in a pair of identical twins?

"Nonidentical identical twins" constitute a "disjunction mosaic," an extremely rare phenomenon, because it requires a combination of two highly improbable events: division of the ovum and a mutation.

Genetic differentiation, which creates individual diversity within the species, has far-reaching consequences. In the first place, it has the advantage of reducing the harmful effects of changing circumstances. If a population were made up entirely of genetically identical individuals, they would all be in danger of perishing from an adverse change of environment. But because of genetic diversity, some individuals will have a greater chance of surviving and producing offspring better adapted to the new circumstances.

Some evolution theorists have gone so far as to say that sexual reproduction developed precisely because it creates the individual diversity that is advantageous to the species. But this diversity also has its drawbacks. For one thing, it accounts for the low rate of success in attempts to transplant an organ or tissue from one individual to another. It is as if each individual, each biological person—each "organic homeland," to use the expression of the great physiologist Paul Bert—refused to adopt cellular material from another "homeland."

There is a great difference between the results of an autograft, that is, a graft of tissue taken from the same individual, and those of a homograft, that is, a graft from one individual to another. If a strip of skin is taken from an individual's thigh and transplanted to his back or forehead, the operation is almost certain to be successful. A severed segment of an ear or nose can also be grafted back in place, provided it is done without delay. Homografts, however, are usually unsuccessful. Transplanted from one individual to another, a strip of skin, or an organ, promptly necrotizes and is finally eliminated. Peter's organism opposes Paul's tissues; it struggles against them, displaying a kind of biological xenophobia.

If we recall what has been said about identical twins—who are "the same individual published in two copies"—it is easy to understand why a graft from one of them to the other is quite likely to succeed, because in this case a homograft amounts to an autograft. This fact can be used in a test for determining whether two twins are identical or fraternal; if a small piece of skin from one can be successfully grafted onto the other, we may conclude that they are identical twins.

Identical twins therefore have an added measure of security, because each can supply the other with tissues or an organ if it should be necessary. There have been such examples as that of the man who flew across the Atlantic to bring his seriously burned twin brother the patches of skin that no one else in the world could have given him.

Science is now developing means of overcoming the organism's resistance to homografts. First, it has been discovered that if the grafted material comes from a very young organism, or, better still, from an embryo, it will sometimes be accepted. This principle has already been used successfully in a number of operations. May and Huignard have reported the case of a retarded boy who, after a transplant of parathyroid glands from a newborn infant, grew several centimeters taller and made pronounced intellectual progress.

Furthermore, Peter Medawar has shown, in a series of

masterful experiments that earned him a Nobel Prize, that very young organisms do not reject foreign tissues and that one can take advantage of this tolerance to accustom them to such tissues, so that homografts can later be performed successfully. Thus, if a newborn human baby is given injections of white corpuscles from his father, he will always be able to accept tissue grafts from his father, even in adulthood.

Finally, to overcome organic xenophobia, that is, to favor the success of homografts, immunological resistance, which has its seat in the bone marrow, can be annihilated or temporarily reduced. Penetrating radiation or certain chemical compounds are used for this purpose. By the use of one or another of these methods, science has already created a number of chimeras: individuals whose bodies contain a functioning organ taken from another individual who is not an identical twin. We therefore now have people living among us who, from a genetic point of view, are no longer completely themselves.

In these human chimeras, the transplanted organ or tissue, which is now an integral part of the foreign organism, undergoes no alteration in its genetic makeup, its chromosomes, its nucleic acids. It is not at all "assimilated" by the new organism; it keeps its individuality, its "otherness." Peter's tissues, living in Paul, will not be "Paulized"; Paul's, living in Peter, will not be "Peterized."

Important questions now arise. What happens to the personality of a chimera? Is he depersonalized to any extent, however small, by having within himself an organ that is not his, that has come from outside him? Must we regard a successful homograft as a violation of the biological person? The problem is all the more important, because the number of human chimeras will continue to increase as progress is made in techniques of transplanting and preserving organs.

There is surely little reason to feel that an individual is depersonalized if he has a kidney from someone else, but we are a little more hesitant if he has foreign bone marrow and produces blood that is no longer his own, and still more so if he has a transplanted endocrine gland, because we know that these glands, by their hormones, act on the individual's temperament, disposition, feelings, and emotional reactions. Let us recall Alexis Carrel's famous remark: "One thinks, loves, suffers, and prays with one's whole body."

It can be maintained that once this gland is integrated into another organism and controlled by another nervous system, it will lose its glandular personality. But the problem remains open. Lederberg asks, "What is the moral, legal, and psychiatric identity of an artificial chimera?" Etienne Bernard wrote, "Is a person a whole? Does he depend on an organ? If so, which one?" And the great Pascal spoke in similar terms: "A man is a subject that possesses attributes; but if he is to be isolated in one of his parts, shall it be the head, the heart, the stomach, the veins, each portion of the veins, the blood, each humor of the blood?"

The problem of the biological alienation of the person would, of course, become particularly acute if brain transplants were to become feasible, as Maurice Renard imagined in his novel *Le Docteur Lerne Sous-dieu*. As long as it remains impossible to graft nerve tissue, we may believe that the human person is well defended by nature. But who can say where scientific progress will lead? Petar Nichola Martinovitch has already succeeded in transplanting embryonic brains in fowls.

For most physiologists, the brain is the seat of the person. If, says Paul Chauchard, we could separate the brain from the body, "it is certain that personality would remain with the brain, not with the body, for the brain, the organ of integration and personalization, preserves within its structures the memories that are the basis of the self."

The philosopher Raymond Ruyer holds the same opinion. He places the essence of the person in the brain and the germ cells; all other parts of the body are only auxiliary organs of support or nutrition, which can theoretically be replaced by artificial devices or imitated by automatons.

While Ruyer, Chauchard, and many others reduce the essence of the person to the brain, a legal philosopher, Aurel David, does his best to convince us, with as much skill as passion, that the real person is totally independent of the physical body, which belongs to the realm of property or things. It is a strange, subversive thesis, which calls for comment.

David points out that cybernetics is becoming better and better at imitating human organisms, in structure as well as function. Small devices for regulating the heartbeat are already being manufactured, and there is talk of making artificial hearts.

As for transplants, which are being performed with growing success, do they not treat the living organ as an inanimate object, in the same category as false teeth or hair?

If each particular organism can be considered a thing, the same must be true of all the organs taken together, that is, the body. Therefore, since law, morality, and humanism demand that the notion of the person be saved, it is indispensable to dissociate it from the notion of the body.

From this viewpoint, the problem of transplants does not arise as I described it above. I considered the question of whether the individual was depersonalized to some extent by a transplant, but for Aurel David that is not even a possibility, since his postulate is precisely that no such depersonalization can occur. If, therefore, the body can be altered by a transplant, it is foreign to the person, it is only an aggregate of things, a "collection of organs," a "flesh-and-blood doll," a "protoplasmic robot." Our organs, limbs, hands, and eyes undoubtedly belong to us, but in exactly the same way as our shoes, gloves, or eyeglasses: we own them, but we do not consist of them.

From the legal standpoint, moreover—and Aurel David is a good jurist—the separation of the body from the person has always been an obvious fact: "Jacques gives one of his kidneys to Paul. I owe Jacques a hundred francs. After the operation, do I owe ninety-nine francs to Jacques and at least one franc to Paul? Juridical experience is sufficient to answer no; I still owe a hundred francs to Jacques and nothing to Paul. We must conclude that the man to whom I contracted a debt was composed of Jacques minus his kidney. I no more contracted a debt to the kidney that Jacques was 'wearing' that day than I did to his jacket, which he later gave away." The same reasoning could be applied, theoretically at least, to every other part of Jacques' body.

David goes so far as to wonder whether a man's love of a woman should not be freed of his attachment to the "flesh-and-blood doll," the "feminine husk." The question is all the more pertinent because love, as I have said, is the most sensitive test of personality.

"A man is passionately in love with a woman, and the progress of science has made hand transplants possible. For one reason or another, the woman sacrifices her hand to a friend. No legal difficulties will result from this. But will the man's love be divided between Constance and Camille?"

Of course not: it was the person he loved, not her hand, which is only a material possession, even though it may be "a jewel among possessions." He loved Camille's hand only because it was hers, and only as long as it served her, "as we love camellias, because the Lady of the Camellias wore them on her bodice."

If it is not her hand that we love in Camille, is it her face, her eyes, her neck, her bust, her legs, her voice, her hair? No, because all that also belongs to the realm of things. But it will probably be a long time before we have evolved to the point where we are capable of a tenderness sufficiently clear-sighted and disembodied to enable us to prefer the real person of Camille to her corporal equipment. "It will no doubt take us several hundred years to become used to loving Camille and not her hands."

A strange vision of the amorous future proposed to us by this poetic and philosophical jurist!

David even maintains that the mind and the sensitivities of the "heart" are not part of the person, because all that depends on the brain, the sympathetic nervous system, and the endocrine glands, which, being part of the body, are things and therefore cannot be endowed with personality. Thus, Roxane is mistaken when, opposing the beauty of the mind to that of the face, she refuses to love Christian for his "momentary attire" and wants to adore him for what makes him truly himself. In Aurel David's view, it is no more the mind than the body that makes a person himself; both are equally part of his "attire."

And Don Juan—to remain with Edmond Rostand's heroes —actually has no reason to be disappointed when he learns from his mistresses that they have loved him only for an odor of "light tobacco, the bedroom, and the fencing school."

What, then, is that real person, that central person in favor of whom David repudiates both the physical and the mental self? Is it the soul of the spiritualists? No. It is a mysterious little flame that is probably the same in all human beings. And so, paradoxically, the human person is characterized by his impersonality.

It is corporal machines, says David, that are unique, that differ from individual to individual. The law proclaims "the equality of persons, whether they have blue or green eyes."

At the beginning of this study I strongly stressed, in the name of biology, the uniqueness of the person. As we have seen, David rejects that notion; and one is, of course, tempted to ask him why, if all persons are alike, anyone should prefer Constance's person to Camille's, or vice versa.

Before leaving Aurel David and his strange "personalism," let us note that the great Pascal raised questions concerning the person that are rather close to those raised by our jurist. (It is not by accident that I have just referred to Pascal for the third or fourth time: the author of the *Pensées* was haunted by the problem of the person, a fact that has not been sufficiently emphasized.)

Let us listen to him: "A man stands at his window to watch the passers-by; if I am one of them, can I say that he stood there to see me? No, because he is not thinking of me in particular. But if a man loves a woman for her beauty, does he love her? No, because if smallpox destroys her beauty he will no longer love her. And if I am loved for my judgment or my memory, am I loved? No, because I can lose those qualities without losing myself. But where, then, is that self if it is in neither the body nor the soul? . . . Thus we never love anyone, but only qualities. . . . Let us therefore not make fun of those who make others honor them for their functions and offices, because we never love anyone except for borrowed qualities."

The truth is that the line of demarcation is rather blurred between the real self, the real person, and everything that is borrowed or added. What does it mean to love someone for himself? In the case of a woman, how are we to disregard her hair style, her clothes, her ornaments, her perfume? Or, in the case of a man, his position in society, his renown, or even the kind of car he drives? Furthermore, is there not a little of the self in all that nonself?

Yet despite Pascal's objections and Aurel David's ingenious sophisms, I believe that there is no other human reality than the body we see and touch, that "protoplasmic robot," that "flesh-and-blood doll," that "collection of organs," that "corporal machine"—that physical person, in short, so debatable, so equivocal, so ambiguous, so imperiled, so badly protected, so imperfectly separated from the world of things.

Of course, I share Aurel David's concern to some extent; like him, I am somewhat perturbed by the idea that the human person, so sacred to me, should be capable of being divided, taken apart, fragmented, and partially replaced, manufactured, and imitated. How can I deny it? Whether we like it or not, we must become more and more accustomed to seeing the person treated as a thing by science and technology, because the available means of adulterating and rectifying the person will become increasingly effective.

Vitallium skeletons, silicone tracheas, plastic corneas, metal cardiac valves—and we are, of course, only at the beginning of this "thingification" of the human body. Need I also mention hormone treatments, brain surgery (which has been called "personality surgery"), and the whole disquieting pharmacopoeia of "psychochemistry"?

All this is certainly admirable; and if one of those means can save or prolong the life of someone we love, our philosophical objections will not carry much weight against the hope of being able to spend a little more time with that person the exact qualities of whom we find it increasingly difficult to define, but whose mysterious reality becomes overwhelmingly important to us as soon as we are threatened with losing it. In our more detached moments, however, we feel a strange uneasiness at seeing science intrude to such an extent into the intimacy of the physical and moral person.

How far will it go in that direction? Tomorrow, perhaps, fatigue, anxiety, and grief will be abolished. Sorrows will be deadened like toothaches. Pleasures, joy, and happiness will be dispensed chemically. Feelings, opinions, and ideas will be controlled. Certain memories will be removed and replaced with others; even the past will be falsified.

Tomorrow, not content with acting on the body, science will

act directly on the germ cells, modifying the person at his source by changing the composition of the nucleic acids that determine heredity. Tomorrow, people will be reproduced by a method analogous to the propagation of plants by cuttings, and, without regard for the narcissism of those mass-produced human beings, technicians will turn out as many copies of an exceptional person as may be desired.

Slight and superficial though they may be, we cannot ignore the falsifications that are already being inflicted on the physical person by plastic surgery and other cosmetic techniques. Hair dyes, permanent waves, false eyelashes (Parisian women reportedly buy eighteen thousand pairs of them each year!), operations to change the shape of the nose (we see the faces of television stars "mutating" from one week to the next), contact lenses that change the color of the eyes.

What would La Bruyère say today, having condemned makeup in his own time? "It is a kind of lying," he wrote, "to seek to deceive the eyes of others and have an external appearance that is contrary to truth."

Let us admit that we no longer know very well *whom* we look at, whom we admire, whom we love. Before those manufactured splendors, those masterpieces of surgical art that devalue natural beauty (the other day I heard a young woman say, "It doesn't do you any good to be pretty nowadays!"), we are reminded of the sorcerer in Villiers de l'Isle-Adam's novel *L'Eve future*, who preferred his own artifice to that of others and therefore made a whole woman for himself.

Having noted the ways in which the organic person is already being adulterated and is likely to be adulterated still more in the future, I must now go on to the causes of *moral* depersonalization that seem to be inherent in our time.

The growing importance of machines, the regimentation and standardization of activities, the increasing control exercised by government bureaucracies that are becoming more and more indiscreet and meddlesome—everything combines to devalue the individual, to frustrate his need for distinctness, to humiliate him in his narcissism, to engulf him in a mass in which he feels powerless, anonymous, and despised. The whole universe constituted by a human being is reduced to a number, a card in a file, an abstraction. "Next, please," sings Jacques Brel, and that could well serve as the dreary refrain of our banal and indistinct existences.

Not to mention the constantly refined techniques of propaganda, which, by distributing the same "governmental truth" to everyone, tend to make minds uniform and keep them in subjection. Considering how crowds are turning more and more into herds, and seeing, wherever we look, how men are subjugated, conditioned, trained, and blended into a homogenous mass, we cannot help wondering anxiously what fate lies in store for the human person. Will a kind of spiritual totalitarianism not eventually destroy that fragile "category of the self" that, as Mauss said, has so slowly "grown through many centuries and vicissitudes"?

Speaking of Aldous Huxley and his chilling *Brave New World*, Raymond Las Vergnas has evoked the danger of all-inclusive planning, which, on the pretext of organizing the human ant hill, reduces individuals who were previously unique and irreplaceable, to "interchangeable cogs in a senseless machine. . . . Take warning," he concludes, "because tomorrow it will be too late. And even today it is already very late."

Yet there are admirable moralists who, with steadfast optimism, refuse to share the anxiety that the future of the human person arouses in so many good minds. Whether secular thinkers like Marie Jean Guyau or Christians like Teilhard de Chardin, they deny that the course of our civilization is necessarily contrary to the true interests of the individual. While they admit that the tightening of social bonds, the increasing scope of communications, and the "fusion of sensitivities" may sometimes have a limiting or even oppressive effect on people, they reject the idea that there is any essential antagonism between the element and the whole, between the individual and the group, between the personal and the universal.

"The irreversible process that assembles us into some sort of vast organic unity," writes Teilhard de Chardin, "should exalt our personality, rather than impairing it," because true unity, far from indiscriminately mingling those it brings together, accentuates their differences, makes their originality stand out, "ultrapersonalizes" them.

To accede to the plural without denying the singular, to be enriched by others without being emptied of oneself, to achieve harmony without conformity, concord without unison —that is surely the ideal toward which we must strive, and a society deserves our trust and devotion to the extent that it enables us to approach that ideal.

Whatever the future of man may be, whatever direction his progress may take, and whatever gains he may pride himself on making in power, efficiency, knowledge, or even happiness, the cost of all this will be too high if it must be paid for by a permanent diminution of the human person. As long as it is still permissible for us to form and express a personal opinion, let us hasten to proclaim that we prefer a discontented human race to a herd of satisfied "rhinoceroses."

THE LIMITS OF
THE HUMAN

In *Zoo, ou l'Assassin philanthrope*, a curious play that was performed in Paris with great success, the writer Vercors, author of the unforgettable *Silence de la mer*, imagines that a scientific expedition to some still unexplored part of the earth has discovered a species of anthropoids close enough to the human race to raise doubts as to whether they are animals or men.

A female "tropi" (the name given to the species) is impregnated with human semen and gives birth. Her offspring is taken to London, and his birth is registered; soon afterward, his father kills him with an injection of strychnine.

This raises a legal and moral problem: in killing the offspring of a female "tropi" and a man, has the father committed a crime against society and the human conscience? Does the life of a "tropi's" child deserve enough respect to be protected by our laws?

The problem is discussed by anatomists and zoologists, who

meticulously compare the structure of "tropis" with that of men. Many other arguments could be added to those exchanged by Vercors' characters. If, today, we really had to situate a newly discovered species in relation to man, we could use not only morphological criteria, but also physiological, immunological, cytological, and biochemical tests; we could compare the two species' karyotypes, that is, their complements of chromosomes; we could perform grafts to see whether tissues from one would be tolerated by the other, and so on.

But whatever the arguments that might be used by scientists to determine the degree of affinity between the two species, Vercors rejects them in advance, and that is precisely where the deepest meaning of his work lies. What he wants to make us understand is that the criterion of humanity cannot and should not be a zoological criterion; he wants to convince us that the dignity of being a man is independent of structure, external form, and even inner organization.

"I am trying to show," he writes in an article (*Bref*, January 1964), "that a creature with a human or even an angelic appearance, but with strictly animal behavior, would not be entitled to that dignity, whereas another, who had a simian appearance, or worse, but who, like us, was aware of his condition, refused to submit to it, and rebelled against its laws, would be entitled to the same dignity as ourselves, since this quality stems from that resistance and rebellion."

Here we recognize one of the favorite themes of this great Resistance fighter, author of *Les Animaux dénaturés* and *Plus ou moins homme*. According to Vercors, as soon as we let ourselves be drawn into the area of morphology, grant importance to structural appearance, or consent to take physical analysis into account, we "open the door to racist contradiction." If we accept the idea that a living being is not a man, or at least that he is "less human," because of certain features of his skull or thighbone, others may later use that concession as grounds for excluding a human group from the human race, or at least denying it a full share of humanity.

Do "tropis" have human behavior and awareness? If so, that is enough to make them human, even if they have four hands and a prehensile tail. It is useless to examine their skeleton, measure their facial angle, make a plaster cast of their heel bone, or determine whether their cranial capacity reaches that volume of eight hundred cubic centimeters that Vallois calls the "cerebral Rubicon"; it is useless to scrutinize their brow ridges, sinuses, condyles, or sutures. Humanity is not defined by the body, but by the mind.

For Vercors, as we have seen, the specifically human mentality is characterized by a secession from nature, an attitude of insubordination; in this he agrees with the philosopher Edouard Le Roy, who defined man by the capacity for revolt. Others have defined him differently: by an aptitude for articulate language, by toolmaking, by the feeling of existential anxiety, by "psychic distance," by the ability to live in the context of time, by a sense of responsibility, by a combination of guile and comradeship. Personally, I doubt that any single trait is sufficient to express the essence of humanity; and I confess my fear that efforts to enclose humanity in a word may lead to a kind of psychological racism as dangerous as any other. What a temptation for the fanatic—and there are always fanatics everywhere—to think that his adversary is less human than himself because he lacks some mental or spiritual quality.

I do, however, share Vercors' view that the psychic must be given priority over the corporal and that living beings endowed with intelligence and sensitivity would certainly deserve to be treated with consideration and respect, no matter what their physical structure might be.

This is also the opinion of the philosopher Raymond Ruyer. "Let us suppose," he says, "that two kinds of living beings have been discovered in some remote corner of the world. The first show a certain shrewdness and are human in appearance, but have neither a cultural tradition nor a language and are incapable of learning to speak. The second have long tails, but they also have a language and a culture. Biologists may be more dubious of the second than of the first, but ethnologists will not hesitate to consider the long-tailed possessors of a language and a culture as men."

The biologist, of course, cannot help having a certain doubt about the possibility of such an independence, such a disjunction, between body and mind, but he does not reject it categorically. After all, beings who are highly intelligent, though built quite differently from us, may exist on other planets and may someday come to us in flying saucers or some other kind of vehicle.

"Are we," wrote Guyau at the beginning of this century, "the only thinking beings in the universe? We can . . . without too much implausibility, accept the idea that the universe contains a multitude of human races analogous to ours so far as essential faculties are concerned, though perhaps very different as to the form of their organs and superior or inferior in intelligence. These are our planetary brothers. In the dreamland that Fontanelle, Diderot, and Voltaire enjoyed exploring in the past, imagine a human race descended not from anthropoids, but from the animal which, with the monkey, is the most intelligent on our earth: the elephant."

Guyau develops this fantasy with satisfaction, pointing out that "the elephant's trunk is, with the hand, one of the strongest and most delicate prehensile organs of any animal species. Thus, there might have been on our earth itself, or on some faraway planet, a giant civilization quite different from our civilization, in its external aspects, if not in its general laws. We must become familiar with the thought, so repugnant to our instinctive anthropomorphism . . . that the order of dignity among species could be rearranged without suspending the general progress of evolution." (*Irréligion de l'avenir*.)

These ideas of Guyau's are not at all contradicted by modern thought. In anticipation of strange visits that we may receive tomorrow, efforts are now being made to create languages whose extreme generality will enable us to communicate with intelligent beings radically different from us.

Science fiction has gone even further than Guyau, with his sagacious pachyderms; some writers have imagined giant ants endowed with superhuman intelligence, and even flowers capable of thought.

But one remark is called for here. If we readily grant that a being physically different from man can bear within him what we respect in man, it must remain clearly understood, I believe, that anyone who belongs to our species, anyone born of a man and a woman, must be regarded as human and treated as such, even if some hereditary or acquired defect makes him lack the faculties we consider characteristic of man.

Young children deprived of human contact may become permanently "animalized." Should such "wolf children" be treated other than as human beings?

Who does not see the danger in failing to recognize man in man? Biological form in itself commands respect, although it is not everything; it alerts us, predisposes us. If someone seems to have only the outer appearance of a man, perhaps his humanity is merely dormant or hidden; perhaps he has not encountered the conditions necessary to the expression of his human potentialities. In any case, he is one of us.

In his *Gulliver's Travels*, Swift, a fierce misanthrope, endows his imaginary horses with all the virtues he denies to the Yahoos, who are hideous caricatures of ourselves. But I confess that while I respect the beautiful and noble Houyhnhnms, incapable of lying or harming their fellow creatures, I cannot wholly detest the wretched Yahoos. Despite their baseness, bestiality, dirtiness, cruelty, duplicity, and callousness, I recognize them as brothers, members of our tribe, our clan, our family.

Among the ancestors of man there were ambiguous, equivocal creatures, like "tropis" in that if they were to appear before us today we might not be sure whether to treat them as brothers or as big game. But man is now thoroughly isolated in the animal kingdom. There is no longer much talk of Yetis, the famous "abominable snowman"; and those who have carefully studied the anthropoid apes are of the opinion that, with respect to mental faculties, there is an unbridgeable gap between them and us. They are outright animals, and no attempt at education—like that of the Kelloggs, for example, who raised a young chimpanzee in the same way as their own son—has ever resulted in the slightest degree of humanization. There are, as we have seen, conditions capable of dehumanizing a man; but we know of none that can humanize an ape. It is impossible to teach them a language, to communicate with them. Although their intelligence may enable them to make rudimentary tools, it is not capable of improvement.

In the eighteenth century, there was less certainty about this insular situation of man. The philosopher Julien Offroy de La Mettrie, referring to the wonders that Joseph Konrad Amman had performed with his deaf-mutes, stated the conviction that monkeys could be even better pupils, provided they were neither too young nor too old and had lively faces.

"Not only," he said, "do I defy anyone to cite any truly conclusive experiment that would prove my project to be impossible and ridiculous, but there are such structural and functional similarities between us and monkeys that I have little doubt that, if a monkey were perfectly trained, he could eventually be taught to speak and therefore to understand a language. He would then no longer be a wild man or a near-man, but a man in the full sense of the word, a small civilized man with the same abilities and physical equipment as ourselves for thinking and taking advantage of his education."

At about the same time, however, Georges-Louis Buffon made a radical separation between man and the monkey, on the grounds that the former had an immortal soul, a "thinking principle," which the latter lacked.

The orangutang, said Buffon, shows us clearly that the soul, thought, and speech do not depend on the organization of the body, that they are special gifts, given to man alone: "The Creator did not wish to make man's body on a model absolutely different from that of animals, but He breathed a divine spirit into that body. If He had bestowed that same favor, I will not say on the monkey, but on the vilest species, the animal that seems to us the most badly organized, that species would soon have become a rival of man. Vivified by the spirit, it would have surpassed other species; it would have thought, it would have spoken."

Today, oddly enough, it is no longer in the monkey, the ape, or the elephant dear to Guyau that biologists expect to find intellectual abilities close to man's, but in a sea mammal, a cousin of the whales: the bottle-nosed dolphin (*Tarsiops truncatus*).

Ancient stories—notably the one told by Pliny, concerning a child who rode a dolphin to school every day—testified to that animal's intelligence and friendliness, but specialists gave them no credence until 1961, when an American neurophysiologist, John C. Lilly, drew attention to the amazing mental capacities of the dolphin. (See *Man and Dolphin*, Doubleday, 1961.)

Not only does the dolphin have a brain as large as man's, and therefore much larger than that of the anthropoid apes, but he is also able to emit a wide variety of sounds and even to pronounce words of human language. He appears to be capable of such great intellectual progress that no limits can be assigned to it. His quickness to learn, his friendly inclinations toward humans, his docility, and his sense of cooperation would seem to indicate that he can become one of our "valid interlocutors" and even a valuable helper in undersea exploration.

Lilly predicts that in the rather near future we will be able to communicate profitably with dolphins, provided we rid ourselves of the superiority complex that too often characterizes and vitiates our relations with the animal world.

If Lilly's conclusions are well founded (and it must be said that they are still vigorously contested by a number of his colleagues), the dolphin may be one of those creatures whose existence we have already accepted hypothetically: creatures that do not have a human form but are endowed with an intelligence resembling that of man.

Lilly goes so far as to foresee (rather "Vercorsianly") that, if dolphins succeed in attaining the conversational level of a simple-minded man—which is much higher than that of a total idiot—they will pose an ethical, legal, and social problem for man, because they will have crossed the threshold of humanity: "If they go above this level the problem becomes more and more acute and if they reach the conversational abilities of any normal human being, we are in for trouble. Some groups of humans will then step forward in defense of these animals' lives and stop their use in experimentation; they will insist that we treat them as humans and that we give them medical and legal protection." (*Man and Dolphin.*)

A touching anecdote illustrates this growing dignity of the dolphin. In 1956, a dolphin came to the beach at Opanoni, New Zealand, and mingled with the bathers. His playfulness soon won him the friendship of the whole population. When he died (accidentally) an ordinance had just been passed giving him absolute protection. His death was mourned by everyone, and a monument was erected to his memory.

If, for the moment, there are not yet any animals on earth with human or quasi-human faculties, we can still wonder whether science will produce such creatures some day. Let us recall H. G. Wells's famous novel, *The Island of Dr. Moreau,* that fantastic tale in which a surgeon, starting with animals, creates grotesque and pitiful creatures that are speaking beasts, caricatures of humanity, barely above the level of idiots. Torn between their bestiality and their humanity, half rebelling against nature, the "tropis" fashioned by the old English doctor try to follow the law imposed on them by their human creator as best they can. Every evening they recite the litany they have been taught: they must not lap when they drink, they must not walk on all fours, they must not hunt other men, they must not claw the bark of trees. We probably cannot expect surgical modeling, however skillful, to produce such near-men; but there are subtler procedures—notably transplants.

It is permissible to wonder whether very young apes might not become humanized to some degree if, for example, human bone marrow were transplanted to them, so that their bodies would continually manufacture human blood. Professor Georges Mathé, with a view to solving certain problems related to cancer, has already begun working toward the creation of such "chimeras" by incorporating human tissues into an ape.

One might also consider modifying an ape's body fluids either by introducing hormones of human origin or by other chemical medicaments capable of activating the functioning of the brain and thus improving the intellectual faculties. (It might be advisable to inject young apes with ribonucleoprotein molecules taken from a human brain.) Some cases of

hereditary idiocy are cured by diets that modify metabolism—
may it not be that apes, in relation to man, are afflicted with a
kind of genetic idiocy?

It is also permissible to envisage treatments which, when
performed on an ape embryo, would increase the number of
cortical cells in the brain. Experiments of this kind seem to
have given positive results in frogs and even in rats, but they
have not yet been performed on chimpanzees. And finally, the
progress of molecular biology may make it possible to modify
at will the genetic material, the DNA, of the anthropoid apes.

The Reverend Father Riquet said one day, "We will not
allow biologists to attack the problem of 'superhumanization'
until they have made a man from a monkey." This is a disturbing remark because, before achieving complete success,
how many partial successes we must anticipate and fear! How
many ambiguous creatures, wretched "tropis," must come
into being, and what strange gleams will be seen in their eyes.

While it was believed in the past that apes were only a kind of
man, it was also believed that some men were only a kind of ape.
Europeans had a strong tendency to think of Africans, whether
Kaffirs or Hottentots, what Buffon thought of orangutangs—
that those "savages" had no souls.

In 1550, Juan Ginés de Sepúlveda said that the natives of
America lacked reason and were "as different from Spaniards
as cruel men are from gentle men, and monkeys are from
men."

In the eighteenth century, the philosopher Jean-Baptiste-
René Robinet wrote, "It has been said that the orangutang is
an animal behind a human mask; it could also be said that a
Hottentot is a man disguised behind the features, the voice,
and the behavior of an animal." And in a history of Jamaica
written by a certain Mr. Long, the author expressed the view
that, although Negroes could not be regarded as totally unfit

for civilization, since even monkeys could be taught to eat, drink, and dress like men, they were of all known human species the least capable of thinking and acting like men (barring a miraculous intervention by Providence), because of the natural baseness of their minds. He added that he did not feel it would be dishonorable for a Hottentot woman to have an orangutang as her husband. It is hard to believe that the good Mr. Long was not writing with a pen dipped in the black humor that seasons the avenging irony of Montesquieu.

It is always a pleasure to quote Montesquieu's caustic lines on Negro slavery (*Spirit of the Laws*, Bk. XV, Ch. V): "Those of whom we are speaking are black from head to foot, and they have such flat noses that it is nearly impossible to pity them." And: "It is impossible for us to suppose that those people are human, for if we did, we would begin to believe that we ourselves were not Christians." We know the result of the belief that "color constitutes the essence of humanity": men treated like animals or, worse still, like objects, things, merchandise.

In the middle of the last century an eminent scientist, Professor Louis-Pierre Gratiolet (in *Mémoire sur les plis cérébraux de l'homme et des primates*), accepted a hierarchy among the races, with the lowest ones representing a transition from the ape. He did not conclude, however—for he was a good-hearted anatomist, not a slavetrader—that those inferior races could be treated with complete callousness: "The law of humanity, which protects and gives maternal care to the most monstrous idiots and the most degraded cretins, extends to all the human races. We have no right to do violence to them, or lie to them, or kill them. With regard to the weak, we have only the right of charity."

Even today, not everyone has such a benign variety of racism. There are still places on earth where odious discriminations are practiced, and segregations that dishonor the segregators. Even allowing for the exaggeration that characterizes political demonstrations, it is significant that not long ago white people in a Chicago suburb shouted to blacks, "Go back to your trees, you monkeys!"

It is important, however, to point out that science, at least, is no longer an accomplice in such repulsive manifestations of racism. Anthropology, better informed, now teaches us that belief in a racial hierarchy cannot be justified by any proven facts. People do differ according to their race: they differ physically, perhaps mentally also, and it is not impossible that some races might be genetically better adapted than others to certain forms of civilization; but such a difference would imply no inequality. From that standpoint, there is no problem concerning the limits of the human; all men are fully human.

This leads me to say a few words about *Lettre sur les Chimpanzés* (1965), by Clément Rosset, a clever little pamphlet written with Swiftian humor. In it, the author pretends to be magnanimously pleading the cause of chimpanzees, an oppressed race scandalously misunderstood and subjugated by men; they are the most disinherited of all African populations, and are as justified as any other in asserting their rights and demanding a fair improvement in their standard of living. This irony is, of course, directed against antiracists like myself; my name, in fact, appears in the pamphlet, along with those of Jean-Paul Sartre and Teilhard de Chardin.

The chimpanzee, says Rosset, shares our biological nature and is moving toward our spirituality; it would be pure barbarism to thwart his ascent: "It is still common to believe that chimpanzees are foolish, as it was long common to believe that certain equatorial populations lacked intelligence." But if the chimpanzee stubbornly remains what he is, if he appears to be incapable of progress, it is our fault. We keep him in a position of inferiority, in servitude. Even the most liberal of us give in to shameful prejudice against him. "How few of our young women would consent without reluctance to choose a husband from among the chimpanzees!"

As a result of being kept at a distance and regarded as an inferior, the chimpanzee has finally come to feel like a chimpanzee; he has posited and structured himself as a chimpanzee. He has, in short, "chimpanzeeified" himself. But all this must change. The chimpanzee must now be allowed to take part in the great movement of liberation that is shaking the planet. We must reject the old bourgeois humanism and replace it with a broadened "primatism" that will enable us to recuperate a whole family of fellow creatures. We must realize the illusory and outmoded nature of certain exclusions that have no other basis than "a prejudice of selfishness and pride."

This is enough to show the tone and tenor of Rosset's pamphlet and the ideas behind it. It is meant to be amusing, and there is no reason we should not find it so, provided there is no shadow of misunderstanding in the smile we give to it, provided we take Rosset's pleasantries only as a game, and provided we give no consideration whatever to the idea—which would be quite simply monstrous and stupid—that there could be any connection between those who oppose all racial discrimination and those who demand the right to vote for chimpanzees.

Despite Rosset's verbal agility, apes are only apes, and all men are men.

I have already mentioned some of the ways in which humanity might exist elsewhere than in man. We must now examine the question of whether man can be deprived of humanity, and if so, whether he still deserves all the respect owed to that quality. The great biologist Alexis Carrel wondered "if certain creatures born of a man and a woman really have a human personality" and, for example, "if an idiot whose mental activities are far inferior to those of a dog should be regarded as a genuine person."

He went so far as to propose eliminating such deficient individuals, because, he said, "modern society must be organized in relation to the healthy individual. Philosophical systems and sentimental prejudice must give way before that necessity. Civilized nations make a naïve effort to preserve useless and harmful individuals. The abnormal hinder the development of the normal. This problem must be faced squarely. Why should society not dispose of criminals and madmen in a more economical manner?" Carrel envisaged nothing less than the gas chamber for the elimination of those judged to be incurably insane. This way of thinking is shared by two German authors, Karl Binding and Alfred Hoche, who demand for society "the right to suppress life that does not deserve to be lived."

However offensive they may be to our sensitivities, such ideas sometimes find a few defenders among those who, having frequented the infernal places where victims of hydrocephalism, microcephalism, and other serious forms of retardation are herded together, have learned the extremes to which human debasement can go. And even a Catholic philosopher like René Poirier agrees that in certain incurable cases "the sacredness of the human person poses terrible problems of conscience."

Most doctors, however, remain resolutely faithful to the principles of the inviolability of human life. "There are many cases," writes François Lhermitte, "from deep insanity to anencephalia, where life persists in the absence of all human attributes, and we cannot refuse to help it."

For a doctor, the problem of conscience is perhaps still more acute in the case of a patient who has been "decerebrated" by a prolonged coma or an irreparable lesion of the higher brain centers. Life can be maintained for a long time in such patients by the use of very complex techniques that require the attendance of trained personnel. Is it the doctor's duty to continue doing everything in his power to assure the artificial survival of a patient who, properly speaking, is no longer a human being, if it is true, as Ruyer says, that man is only an animated brain? Is it his duty to prolong an existence that has been reduced entirely to manifestations of animal life, or even vegetable life? Is it not true that the efforts expended and the means employed are out of proportion to the result obtained?

In his debatable but important book, *La Mort a changé*, Alfred Fabre-Luce writes that these near-corpses "are treated as men when they have become machines. Furthermore, other men are implicitly sacrificed to them when medical teams and equipment that could be better employed are immobilized around them. Is it reasonable to place valuable specialists in the service of living corpses? Is it reasonable that a man whose brain has been destroyed should be surrounded by masked attendants who try to shield him from all danger of additional disease? Obviously not." (Some doctors demand the right to take organs for transplants from these "living corpses.")

Yet even in this limiting case the medical spirit persists in its vocation of saving life as long as there is hope, as long as "motor manifestations show the functional survival of a living nervous system, as long as vegetative life continues: these patients in such pitiful condition live for months, even years; we cannot consider failing to give them the care that their condition necessitates." (François Lhermitte.)

It should be noted that the Catholic church is less absolutist in its respect for life, since it maintains that man must not be completely identified with his biological destiny; a decision to discontinue efforts to revive a patient may be permissible when, all relations with the external world having been cut off, "one may assume that the person has disappeared." (Reverend Father Durand.)

Can anyone fail to see the seriousness and immense difficulty of these problems? That difficulty is increased still more by the fact that, in some cases, the progress of medical technology has now made recovery possible, but at such a high cost that, as Professor Hamburger says, "the day comes when we must decide who will be allowed to die." (*Progrès de la médecine et responsabilités*, Second International Congress on Medical Morality, May 1966.)

If a doctor is faced with the "heartrending obligation" of choosing between two lives, will he not be inclined to give preference to the life that is less damaged, less degraded? And if he does so, has he shown any disrespect for life? Is he not, rather, acting on the basis of a more enlightened respect for it?

But there is a moral risk, and almost a defeat of the medical spirit, in beginning to accept the notion of *biological unworthiness*. Who can be entrusted to determine the threshold of integrity, the physical or psychic minimum beyond which the right to live is attenuated or lost? If we assent to the idea that a human being may be killed, or at least allowed to die, we are in danger of drifting farther along the same path. "Once we begin," says Paul Chauchard, "where do we stop? In the interest even of those who are in good health, respect for human life must be absolute."

Not to mention the conscious or unconscious motives that might weight the fateful decision and influence the terrible diagnosis of "dehumanization"; not to mention the self-interest or cupidity of those who might be hampered or deprived by the continued existence of a "living corpse." We do not live among angels.

And, as Robert Spaemann asks, if permission to live had to be given by the state, would not everyone have a painful feeling of insecurity, abandonment, and isolation?

The same problem, involving the same legal and moral issues, arises with regard to the elimination of babies with serious birth defects. Here, the right not to respect life absolutely has been demanded all the more vigorously because the act of homicide is directed against someone who can scarcely be said to exist, someone who still has no clear self-awareness or spiritual personality.

Let us recall the tragic trial in Liège, Belgium, which ended with the acquittal of a mother who, with the aid of a doctor, had killed her seriously malformed baby. The malformation, known as phocomelia, consisted of the complete atrophy of the arms and legs. It had been caused by the mother's use of the sedative thalidomide during her pregnancy. The doctor who had been her accomplice in the infanticide was also acquitted.

There were many people who, in the name of a supposedly superior charity, justifying Nietzsche by the Gospels, approved of the jury's verdict and praised the mother for having had the excruciating courage to snuff out the beginning of a life doomed to distress and unworthy of being lived. Others, however, condemned the trial as a parody of justice that "lowered man to the level of an animal" (Dr. Hindermeyer) and made the state give tacit assent to a murder (Spaemann). Radio Vatican declared that "a wound has been inflicted on the most sacred and inviolable of all rights: the right to life." It is true that one is inclined to excuse the mother's criminal act to some extent; but, for the sake of principle, should she not have been convicted, then perhaps pardoned later?

Defenders of the right to commit such murders justify their position by referring to the suffering that lies in store for an abnormal child, the unwholesome and painful spectacle he offers to other children, and so on. But these honorable, humane feelings may be mingled with others that are less noble: cowardice of the parents, wounded vanity.

It must not be forgotten that malformed babies of the type

we are discussing have normal brains and are therefore fully human. Some of them—like Denise Legrix, author of the deeply moving book *Née comme ça*—have succeeded, by sheer courage, in overcoming their infirmities and in leading active, productive, almost happy lives.

And then, as in the case of incurable illness, there is the terrible danger of going further. If we believe that we have a right or even a duty to kill a baby afflicted with phocomelia, how long will it be before we feel justified in killing less seriously malformed babies? "Who should be sacrificed? What are the limits of the normal man?" asks Georges Duhamel. And, as Chauchard says, "Each of us is someone's monster."

It may be that the problem of malformed babies is one of those which, having no perfectly satisfying and reassuring solution, belong to the realm of what has been called "the moral indeterminate."

At a recent convention of Catholic intellectuals, Morvan Lebesque said, "After centuries of morality, we still cannot answer questions like those raised by the trial in Liège: Should malformed babies be killed? Where does man begin?" To which Father Jolif replied, "No one knows what man is any longer."

In any case, it is interesting to note that, although the elimination of malformed babies was apparently a common practice in ancient Greece and Rome, and although Auguste Leyser wrote in 1778 that "monsters should be sacrificed with impunity," jurists and moralists began having serious differences of opinion on the subject in the nineteenth century.

For Raubert (1836), killing a "monster" did not constitute homicide, since such a creature had "no more personality than a corpse." But for Erschback, a professor at the Strasbourg law school (1847), "Everyone who has come from a woman's womb is human, and therefore inviolable."

For others, the "monster's" right to life depended on the seriousness of his malformation and particularly on whether his face had a human or a bestial appearance, since an abnormal condition of the organs was not enough in itself to exclude anyone from humanity. Always the terrible problem of the "limits of the human"; they are not easy to delineate.

Dr. Martin, from whom I have borrowed these various quotations, was convinced that no European court of law would ever "fail to punish the crime of killing a malformed baby if criminal intent is clearly established." Those words were written in 1880; it must be acknowledged that, since the Liège trial, respect for the lives of malformed babies has been declining in our civilization.

The problem of malformed babies brings us to that of embryos, which for one reason or another—certain medical treatments of the mother during pregnancy, or certain infectious diseases, such as German measles—will probably be deformed. Some people feel that abortion is justified in this case; others feel that a probability of malformation, no matter how great that probability may be, is not sufficient justification; and still others feel that even a certainty of malformation would not make abortion acceptable. (Certainty might be obtained from X-ray photographs of the fetus.) Thus, there are all degrees of respect for life in the case of an embryo.

René Poirier concedes that abortion may be justified when malformation is nearly certain. But what does "nearly certain" mean? Is it ninety-nine percent certain? Ninety-five percent? Ninety? One thing is clear: here we are on shifting ground.

Not only has the right to eliminate a malformed or presumably malformed embryo been demanded, but also the right to eliminate a normal embryo. For the Catholic conscience there is no problem: the human being is inviolable from the moment of conception, that is, as soon as the ovum has been fertilized. According to Abbé François Maire, there is no essential difference between a fertilized ovum, which is simply an unborn child, and a child that has already been born: "Both are human beings."

And biology recognizes that the human being exists potentially in the fertilized ovum, since it is strongly structured and already individualized, personalized.

Another extreme opinion, in the opposite direction, grants the mother the right of life and death over the fetus in her womb, because it is part of her. "It has no name, no face, no independent existence, whereas a newborn baby is another person." (François Giroud, *Express*, August 25, 1962.)

Whether we like it or not, the respect given to an embryo or fetus varies according to its age and the stage of development it has reached. Here again, we encounter certain difficulties in trying to fix the limits of the human. Shall respect for life be measured by the weight of the embryo? By the number of cells it contains? Shall it be considered wrong to kill an embryo when it has attached itself to the uterine wall? When it has begun to develop a human shape? When the features of the face appear?

Although the penal code recognizes the existence of the embryo from the moment of conception, and theoretically protects it from that time onward, the collective conscience accepts different degrees in the human worth of the embryo. If the seriousness of a crime is to be judged by the number of individuals capable of committing it, it is certain that women capable of eliminating a fetus are much more numerous than those capable of killing a newborn baby and that still more women are capable of destroying a very young embryo, with no more qualms than if they were drowning a kitten. This may be partly due to lack of imagination, or even to lack of information. It is not generally known that a fetus only a few weeks old is already a well-constituted little human being.

Some legal codes—in Eastern European countries, for example—define the legitimacy or illegitimacy of an abortion according to whether it takes place before or after the third month of pregnancy. (The French code of medical ethics stipulates that abortion is permissible only insofar as continuation of pregnancy would endanger the patient's health.)

To a certain extent, even doctors attribute limited worth to a fetus, since they feel that the mother should not be sacrificed to the child, "because she is a human being whose inner life has acquired a definite personality, which the fetus has only in a state of potentiality." (Hamburger.)

It should also be noted that Christians, or at least some of them, grant the right of abortion to a girl who has been raped, "because the unfortunate girl should not be forced to bear a child that would remind her of the crime committed against her." (de Pury.)

It is well known that many illegal abortions are performed in France each year: between eight hundred thousand and a million, about the same as the number of births. The causes of these prenatal murders are often economic and social: inadequate salaries, the housing shortage, or, at the upper end of the social scale, bourgeois respectability. How many fetuses have been sacrificed to family honor!

There can be no doubt that the number of these crimes would be greatly diminished if means of contraception were made more available. The abortion rate is much lower in countries, such as the United States, where contraception is a common practice. But it must be pointed out that, among the birth-control methods that have been proposed, there are some that consist not of suppressing the production of ovules, but of preventing the fertilized ovum, or rather the young embryo, from attaching itself to the uterine wall. In short, however we may feel about it, their effect is to cause a very early abortion.

It may be only a minuscule crime to kill a human being a

few days old and a few millimeters long, with no specifically human form. But it is still a crime, and one that may not be easy to reconcile with respect for human life.

Experimentation on human embryos poses problems of conscience. In some laboratories human embryos are used in research. This is not a crime, properly speaking, since the embryos used are not viable; but is it possible to retain respect for humanity and still accept that rather casual way of handling those "products of a man and a woman" as if they were little animals? It has even been suggested that human embryos might be used as a source of therapeutic substances, if it were proved that such substances could be taken from them. Would it be morally admissible to produce and cultivate human embryos on a mass-production basis for utilitarian purposes? Paul Valéry considered a similar problem: what would happen if it were discovered that a cure for cancer could be obtained from newborn babies?

We must now approach a very different question, one that is fiercely debated: that of capital punishment. If a man has committed a certain antisocial act, does he lose his humanity to the point where he can rightfully be declared unworthy of living and killed like a harmful animal? In other words, if a man is regarded as a "moral monster," does society have a right or even a duty to kill him?

Plato felt that a criminal was still a human being. But in the eighteenth century Pierre Louis Moreau de Maupertuis, who was by no means a bloodthirsty man, saw no objection to practicing vivisection on criminals for the advancement of science: "Perhaps we could make many discoveries about the marvelous union of the soul and the body if we dared to seek its bonds in the brain of a living man. One should not be swayed by any imagined air of cruelty here; a man is nothing, compared with the human race; a criminal is even less than nothing."

For Maupertuis, a criminal did not even deserve as much respect as a useful animal: "Only harmful animals may be treated like murderers or robbers." And I am willing to bet that even today there are many people who would be more inclined to feel pity for a good dog than for a strangler.

The arguments used by supporters of capital punishment are familiar to everyone. A murderer does not deserve to live. Even if he were imprisoned for life, he would still constitute a danger, and also be a burden on the state. Finally, and above all, capital punishment serves as an example: it deters a certain number of potential murderers and thus protects the lives of the innocent.

Actually, however, statistics show that capital punishment does not have the deterrent effect often attributed to it. The murder rate has not increased in the many countries where the death penalty has been abolished, that is, in all European countries except France.

Hence the apparently well-founded conclusion of those who wish to abolish capital punishment. If the death penalty is not an effective means of preventing crime, if there is no reason to fear that innocent victims will have to pay the price of a deluded sentimentality or an ideology that is unconcerned with reality, and if society can decline to answer evil with evil, murder with murder, without weakening itself, why should it fail to set that example?

The abolitionists—and I am one of them—feel that it is always good, whenever it can be done without harming others, to show such a firm, resolute respect for human life that it cannot be shaken by even the most repugnant aspects of humanity. As Paul Chauchard says, "Whether it is a question of malformed babies or criminals, solving the problem by murder means taking the easy way out. Nothing is more contrary to human dignity than the horror of an execution, by whatever means it is performed. More and more, people are coming to see that the death penalty is a survival of the past, a lesser evil that should no longer be tolerated."

Furthermore, the abolitionists point out that capital punishment is not only useless but also necessarily harmful, like all illusory solutions, because it diverts us from seeking the real causes of murder: genetic, psychological, and especially social causes.

Poverty, squalid housing, alcoholism, ostentatious displays of dubiously acquired luxury, exhibitions of violence and eroticism, the publicity given to the most atrocious crimes— these are the things that should be constantly denounced, and it would be more useful to attack them than to relieve ourselves by executing a few wretched criminals.

Is it necessary, finally, to recall how caste prejudice and blind fanaticism have led so many people to deny the humanity of human beings through the centuries? Even today, there are countries where people are held in slavery, bought and sold as merchandise; even today, in countries that claim to be civilized, people are mistreated, persecuted, tortured, and executed because they do not have the same political opinions as their oppressors.

"Racism," Pierre Gascar rightly says, "is not the only basis of selection that is used for rejecting certain human beings, for regarding them as creatures far below the level of animals." But that is too vast a subject to be discussed here.

We have considered some of the moral problems that arise from the difficulty we encounter in laying down the limits of the human, or at least specifying what it is in man that commands respect.

How shall we behave toward people whom sickness has reduced to a level below animality? Toward people whose brains have been irremediably destroyed? Toward malformed babies? Toward those who, by their crimes, have cut themselves off from humanity? Opinions differ on all these grave questions, depending on how determined one is to defend, maintain, save, or spare a life that may appear to be unworthy of being preserved.

It must be said that these differing opinions are held by people who are equally sincere, have the same level of morality, and sometimes even profess comparable philosophical doctrines. Perhaps it is impossible to decide categorically where the truth lies in every case, for we must often choose among values that are all equally deserving of respect and have no common measure.

I admit that I am one of those who carry respect for human life to its extreme, almost to the point of fetishism, as I have surely made clear in these pages. I belong to that category of minds or sensitivities which Alfred Fabre-Luce calls "vitalistic" (an ambiguous term, since it is also employed with another meaning, but I have no objection to it if it is used without misunderstanding). Yes, I am undoubtedly an ardent "vitalist." But of course we all know that such a position cannot be maintained unshakably in all situations.

We have seen that in some cases a total, inalterable respect for human life can lead to an impasse or a contradiction, and that in certain circumstances it can even result in advocating conduct that would be contrary to the true interests of human life. We have seen that even the most determined, intransigent vitalist must bring himself to make concessions when he is forced into the position of having to choose between two lives, one of which is more seriously damaged than the other. There can be no doubt that Fabre-Luce is right when he says (and Dr. Hamburger would agree) that if there are not enough artificial kidneys for all the patients who need them, those that are available should be used for patients with the best chances of recovery. We can only hope that the present period of shortage will be succeeded by an era of technological abundance, in which there will be enough equipment for all patients.

I grant that respect for human life is not an absolute and that those who profess it must not let themselves be blinded by a desire to save a particular life at any cost. But having written that sentence, I am impelled to revise it by adding that although respect for human life is not an absolute, it is still the closest thing to an absolute in our civilization.

The most fervent vitalist is sometimes forced to accept concessions and compromises, but I would at least like to be sure that they will never cease to be accompanied by resistance, reluctance, regret, misgivings, and even a certain remorse.

For my part, I believe that there is no life so degraded, deteriorated, debased, or impoverished that it does not deserve respect and is not worth defending with zeal and conviction. I believe that even if there are good reasons we can give ourselves for sacrificing one life for another, considering the inadequacy of the material means at our disposal, such a sacrifice involves a kind of defeat, because it confirms acceptance of that inadequacy. Above all, I believe that a terrible precedent would be established if we agreed that a life could be allowed to end because it was not worth preserving, since the notion of biological unworthiness, even if carefully circumscribed at first, would soon become broader and less precise. After eliminating what was no longer human, the next step would be to eliminate what was not sufficiently human, and finally nothing would be spared except what fitted a certain ideal concept of humanity.

I have the weakness to believe that it is an honor for a society to desire the expensive luxury of sustaining life for its useless, incompetent, and incurably ill members; I would almost measure a society's degree of civilization by the amount of effort and vigilance it imposes on itself out of pure respect for life. It is noble to struggle unrelentingly to save someone's life, as if he were dear to us, when objectively he has no value and is not even loved by anyone.

I admit that I am a little disconcerted by the spirit of certain books and articles that come my way, when I see that an author is trying to take the tragedy out of death (which, as a moralist once said, is "the only thing greater than the word that names it"), when I read that a person's usefulness to his family or his country should be taken into account in deciding whether or not to prolong his life, when I realize that, tomorrow, any man may have to justify his right to life and prove his organic worthiness. I admit that I am scarcely reassured when Fabre-Luce tells us that "all insane, vitiated, or degraded life is a profanation," and that, in the future, "we shall have to become more and more exacting with regard to the conditions of human worthiness." I am even less reassured when I think that it might not be men like Fabre-Luce who would set the limit of that minimum worthiness. I admit that he frightens me a little (and not only because I myself am in my seventies) when he writes in his brilliant book, *La Mort a changé*, that "for patients beyond the age of ninety, the doctor should be only a helper of death, ready to seize the first opportunity."

I admit that I feel uneasy when I hear someone proclaim "the right of the more alive over the less alive," because that right is a little too much like the right of the fittest or the strongest. I admit that I see in all this a slight tinge of Nietzscheanism that is not to my liking, and I also see it in certain passages of Teilhard de Chardin, where he explains that we must sometimes have the courage to abandon stragglers on the road in order to make swifter progress toward our goal.

Yes, I admit that I would be repelled and saddened if I

witnessed the development of a social ethic in which the value of every life ceased to be infinite, and it was considered perfectly logical and natural for a doctor to discontinue a lifesaving perfusion or decline to revive a newborn baby. I admit that I would dread that excessively rational and realistic society whose advent is now being predicted; that society that would mathematically evaluate the amount of protection and care that each individual deserved, taking into account his age, health, biological fitness, social usefulness, and ability to enjoy life; that society in which everyone would receive only his proper portion of medical assistance, and in which, after having judged a man to be not worth saving, conscientious but anonymous experts would sign an order for his death as calmly as if it were a routine report.

It would be easy, of course—and people less "vitalistic" than I am have done so—to contrast the traditional, rather sanctimonious respect for human life with the savage contempt of it that constantly erupts in our civilization. Highways covered with the blood of accident victims, a monstrous technology developed for military murder, preparations for an abominable nuclear or even bacteriological war—in the face of all that, it may seem ridiculous to care about prolonging the life of a patient with irreparable brain damage or an incurable disease, saving a malformed baby, or sparing a criminal sentenced to death.

"If we think," writes Binding, "of a battlefield covered with thousands of young bodies, or a mine in which an explosion has killed hundreds of good workers, and then think of our institutions in which idiots are given painstaking care, we cannot fail to be deeply disturbed by the horrible contrast between the sacrifice of so many of the best elements of humanity and the zeal devoted to preserving lives that are not only lacking in any positive value, but even have a value that we must judge to be negative."

Granted. But, as Spaemann points out, "It is characteristic

that this appalling contrast does not suggest to Binding that since we take care of idiots, we should not sacrifice healthy young men; on the contrary, he feels that since we sacrifice healthy young men, we should also sacrifice idiots." Why be guided by evil, rather than trying to be guided by good?

It appears that, on the whole, the limits of the human have been gradually expanded in the course of human evolution, and there are those who believe that this evolution has not yet ended. But it may change direction, and the general attitude of the public at the time of the Liège trial should give us food for thought on that subject.

After all, it is not impossible that the human race may take the path already being advocated by the least "vitalistic" among us. It is not impossible that it is they who are right, if being right means thinking as people will think in the future. Perhaps the human race will free itself of some of the taboos in which it now takes pride. And it seems likely that if a movement were ever begun in that direction, changes would be rapid and far-reaching.

If eliminating "monsters" became common practice, lesser defects would come to be considered monstrous. There is only one step from suppression of the horrible to suppression of the undesirable. If it became customary to thin out the ranks of people over ninety, those in their eighties would begin to seem very decrepit, and then those in their seventies. Little by little the collective mentality, the social outlook, would be altered. Any physical or mental impairment would diminish the right to live. Each passing year, each stress, each illness would be felt as an exclusion; the sadness of aging and deteriorating would be combined with a kind of shame at still being there.

Such a society would gain all sorts of advantages: efficiency and productivity would soar; useless work, ugliness, and pain would be greatly diminished. It would undoubtedly offer a more cheerful and agreeable spectacle than our present society. No more madmen in asylums, no more hopeless invalids in hospitals, no more "monsters" in institutions, no more murderers in prison, no more distressingly ugly old people in the streets.

Such a cleansed and purified society would be more dynamic, invigorating, virile, robust, wholesome, and pleasant to look at than our own. Pity would be obsolete, and the idea of gaining insight through suffering would be considered absurd. Waste and inefficiency would be banished, and the normal and the strong would benefit from all the resources formerly devoted to the abnormal and the weak. Such a society would represent a return to the spirit of Sparta and would delight disciples of Nietzsche. But I am not sure that it would still deserve to be called a human society.

THE EVOLUTION

OF GENETICS

As the experimental science of heredity, genetics is little more than a hundred years old, but its origin goes back much further. In ancient Greece and in even earlier societies, the fact that children resembled their parents aroused the curiosity of physicians, naturalists, and philosophers, who tried to explain it with the paltry means at their disposal.

Parmenides, Empedocles, Hippocrates, Aristotle, and Galen each had his own theory of heredity; and throughout the ages there were many men who felt that they had cleared up its mystery, either by resurrecting the naïve views of their predecessors or by adding some new element of their own devising.

While theorists were absorbed in their flights of fancy,

THIS is the text of a lecture delivered at the Sorbonne as part of the symposium on genetics organized by the Maison des Sciences.

more pragmatic men were gathering facts concerning the transmission of certain characteristics in animals or humans. In 1645, Sir Kenelm Digby reported the transmission of an extra thumb on the left hand (polydactylism) from mother to daughter through five generations. (This Digby was a curious man, well versed in sorcery. Having married an extraordinarily beautiful woman, Venetia Anastasia, he set about preserving her youth by putting her on a special diet consisting of young capons that had been fed on vipers. She died in the prime of life.)

Several years later, in 1669, Becker mated a white female pigeon with a black male and observed that the offspring were either entirely black or entirely white; he then found that when he mated these new pigeons with each other, their offspring were also either all black or all white.

In 1683, the famous Dutch microscopist Leeuwenhoek mated a white female rabbit with a dark male, producing dark offspring. He concluded that the germ of life came from the male. In those days biologists were divided into two camps: partisans of the female germ, or ovists, and partisans of the male germ, or animalculists.

Among those who advanced the science of animal genetics in the eighteenth century, we can mention Louis-Jean-Marie Daubenton, Buffon's associate. Having set out to create a breed of sheep with fine wool, he succeeded by practicing methodical selection, that is, by having each generaton sired by rams chosen for their unusually good fleece.

More significant were René Antoine Ferchault de Réaumur's experiments in the hybridization of breeds of fowls, which truly foreshadowed modern genetics: "Let us mate ordinary hens with five-toed cocks," wrote the famous naturalist, "and five-toed hens with ordinary cocks; let us mate ordinary hens with a rumpless cock, or rumpless hens with an ordinary cock. If chicks are born of such matings (and they are, and some of them are capable of reproducing their species), it would seem that we may expect to discover facts that will decide the question at issue, for assuming, as we have done, that the germ exists before the act of coupling, and that the only problem is to determine whether it exists in the male or in the female, the chicks of which we are speaking should show us, by possessing or lacking certain features, whether the germ originally belongs to the male or to the female." (*Art de faire éclore*, Vol. II.)

It was an excellent program. For several years Réaumur did his best to carry it out, taking great care to avoid improper breeding. But he never published his results, and in the second edition of *Art de faire éclore*, he stated that he was glad he had not done so, "because the conclusions that it seemed natural to draw were not as consistently supported by later experiments as might have been expected."

It is not surprising that Réaumur's experiments seemed inconclusive to him, because polydactylism and the absence of a rump are Mendelian traits whose mode of transmission could not have been interpreted with the knowledge of his time.

In about this same period there were some remarkable observations on the transmisson of extra fingers in human beings. Réaumur reported one of them in 1751. He had been informed of it by Godeheu de Riville, Commander of the Knights of Malta and a correspondent of the Academy of Sciences. It concerned a Maltese family in which the abnormality (six fingers and toes) was transmitted by the father, Gratio Kalleia, to three of his four children (two sons and a daughter). The three abnormal children had both normal and abnormal children, while the normal child had only normal children.

In 1752, the geometer Maupertuis described a case of polydactylism transmitted through three generations in the family of a Berlin surgeon, Jacob Ruhe. In this instance, the abnormality had its source in a woman named Elizabeth Hortsmann; she transmitted it to her daughter, who had six children and transmitted it to four of them, one of whom was Jacob Ruhe. Maupertuis regarded the case as a genuine "natural experiment."

"It will be seen from this genealogy, which I have carefully traced," he wrote, "that polydactylism is transmitted by both the father and the mother." (*Lettres*, 1753.)

In 1770, Sauveur François Morand, in a remarkable and richly illustrated treatise, described new cases of polydactylism. Recalling the examples cited above, which suggested to him the unfortunate possibility that "a new species of men" might be formed, he asked "if it would not be proper to oppose marriages between six-fingered men and women. I have perplexed several jurists by asking them about Gratio Kalleia and Elizabeth Hortsmann, and they have not yet answered my question."

Henry Baker commented on the transmission of another hereditary abnormality in 1757. This was the case of a young man, Edward Lambert, known as "the porcupine man" because his whole body was covered with a kind of warty shell. He was presented to the Royal Society of London in 1731; twenty-six years later, he had fathered six children, all boys, who inherited his abnormality. Only one of them was still alive. Baker said that such an abnormal individual could undoubtedly have a line of descendants who would all have the same abnormality and that if the accidental origin of the lineage were forgotten, its members would probably come to be regarded as forming a distinct species within the human race.

Thus Baker, like Morand and Maupertuis—whose ideas prefigured those of modern "mutationists"—envisaged the formation of new races beginning with individual abnormalities. Knowledge of inherited defects was increased by further observations in the last quarter of the eighteenth century: the hereditary nature of colorblindness was established in 1777, that of hemophilia in 1793.

Ideas concerning animal reproduction made little progress until the advent of the cellular theory (1839), and it was a long time before this theory had any noteworthy results in embryology and genetics. We can mention the curious attempts of a Geneva pharmacist, Jean-Antoine Colladon, who about 1820 performed systematic experiments in the cross-breeding of white and gray mice. His work, which introduced into genetics an animal that was later to be used so profitably, was not published, but he presented it to the Geneva Physics Society, and we know of it through the detailed descriptions of it given by the physiologist Edwards and the chemist Jean-Baptiste Dumas, the latter writing in a treatise whose authorship he shared with Prévost (*Annales des Sciences naturelles*). It is not impossible that Mendel was informed of Colladon's experiments and that he was directly or indirectly inspired by them.

In this first half of the nineteenth century, there were also experiments by Girou de Buzareingues with hunting dogs' muzzles and chickens' tails, by Isidore Geoffroy Saint-Hilaire on the transmission of polydactylism and ectromelia in dogs and extra limbs growing on the backs of cows, and so on.

In 1850, the German beekeeper, Johann Dzierzon, who discovered parthenogenesis in honeybees, performed a remarkable experiment with those same insects. He crossed German queen bees with Italian drones and found that a queen engendered by such a cross would produce equal numbers of German-type drones and Italian-type drones.

As Phineas Whiting has pointed out, this was a highly significant result, because Dzierzon had thus shown, in the offspring of the hybrid female, what was later to be known as

Mendelian segregation. The males, having been produced by parthenogenesis, exhibited the potentialities of the female's ovules, and the production of two types of males indicated the formation of two types of ovules: those that had received "Italian-breed" unit characters and those that had received "German-breed" unit characters.

Let us also cite the experiments of Guaita, which were conducted toward the end of the nineteenth century, and especially those of Wilhelm Haacke (1893), in the crossbreeding of white mice and parti-colored mice. Here again, the experimenter noted remarkable facts that might have led him toward Mendelism, but he was unable to grasp their implications.

Medicine also contributed valuable data to the growing science of genetics. Doctors made systematic observations, and the notion of morbid heredity became important in the study of abnormalities, diseases, and predispositions.

There were investigations of the hereditary nature of ectrodactylism, or absence of fingers (Béchet, 1829), deafmuteness (Antoine Portal, 1808), night blindness and day blindness (Cunier), cataracts, extra breasts, harelip, webbed fingers, sternal cleft, dislocation of the lens of the eye, retinitis pigmentosa, the tendency to procreate twins, etc.

Dr. William Sedgwick particularly studied (1861–1863) the "sexual limitation of hereditary maladies," or what would not be called morbid heredity related to sex or the sex chromosome. Dr. Prosper Lucas, in his *Traité de l'Hérédité naturelle*, stressed the transmission of psychic traits, notably a predisposition toward criminal acts. Finally, Dr. Ménière, and especially Dr. Boudin, drew attention to the dangers of consanguinity.

But we must not forget the botanists, whose contribution to pre-Mendelian genetics is by no means negligible. In the eighteenth century, Joseph Kölreuter crossbred tobacco plants, showing the influence of pollen on the characteristics of the offspring. In 1824, Goss and Seton presented a paper to the London Horticultural Society reporting on their experiments with crosses between pea plants with green seeds and those with yellow seeds. The first generation of hybrid plants produced uniformly yellow seeds, but in the second generation there were plants with green seeds, and even plants with both green and yellow seeds. When seeds of different colors were isolated and planted separately, it was found that green seeds produced only plants with green seeds, while yellow seeds produced not only plants with yellow seeds but also plants with green seeds and plants with both yellow and green seeds.

Also noteworthy are studies by Sagaret on the crossbreeding of melons (1826), by Gartner (1849), Lecoq (1862), and Godron (1863) on the daturas, by Naudin on the daturas, the linarias, and the nicotianas (1863), by Wichura on willows (1865), by Charles Darwin on snapdragons (1868), by Vilmorin on the improvement of plants by individual selection of progenitors (1886), etc.

THE CELL NUCLEUS, SEAT OF HEREDITY

Let us see how the problem of heredity stood at the beginning of 1900, the year when the rediscovery of Mendel's Laws was to bring about the emergence of genetics in the true sense of the term. A multitude of facts had been amassed by observation and experiment, but up to that time they had given rise to no overall view, no general theory. Scientists studying heredity had the feeling that they were dealing with phenomena so complex and capricious that there could be no hope of subjecting them to precise analysis. A vast chaos of jumbled facts existed that seemed impossible to sort out and arrange into orderly patterns. The eminent zoologist, Yves Delage, stated this disillusioned axiom: "In the field of heredity, anything is possible, and nothing is certain."

But although examination of the known facts had proved to be so disappointing that the most profound skepticism seemed totally justified, some biologists formulated precise opinions and even constructed ingenious theories concerning the seat, nature, and properties of heredity.

As a result of revelations made by embryologists and cytologists—notably by Edouard Van Beneden in 1883, on fertilization in ascarids—there was a growing tendency to believe that the hereditary material, or idioplasm, as it was then called, was situated in the nucleus of the generative cell, and more precisely in the particles the nucleus contained, the number of which particles being constant for each species. These particles, discovered by Wilhelm Hofmeister in 1848, had been named "chromosomes" by Heinrich Waldeyer in 1888, because of their selective absorption of certain coloring substances. Eduard Strasburger, August Weismann, Karl von Nägeli, and Oskar Hertwig, notably, formulated a chromosome theory of heredity that basically prefigured our modern chromosome theory.

We must stress the major importance of Weismann's truly prophetic work. He is not always given his rightful place in the history of genetics, because he is too often remembered only for his vigorous denial of the transmissibility of acquired traits.

"The fact that, rather than merging with a certain irregularity," he wrote, "the nuclei of the ovum and the seminal cell regularly arrange their ansae [chromosomes] two by two, one facing the other, and thus form a new nucleus, the cleavage nucleus, clearly shows that the organized nuclear substance is the sole agent of hereditary tendencies. . . . The truth is that Van Beneden himself did not draw these conclusions."

"Therefore," he also wrote, "only the nuclear substance can be the vehicle of hereditary tendencies. . . . The essence of heredity is the transmission of a nuclear substance endowed with a specific molecular structure."

Furthermore, there was increasing acceptance of the idea that the hereditary material itself was heterogenous and particular, constituted by a large number of distinct, separable elements, more or less independent of one another. This atomistic or, to use Delage's term, "micromeristic" concept of heredity took widely varied forms. Charles Darwin's "gemmules," Herbert Spencer's "physiological units," Nägeli's and Hertwig's "idioblasts," Hugo De Vries's "pangenes," Haeckel's and Erlsberg's "plastidules," Weismann's "determinants"—such are the principal names given to these particles, whose properties and functions varied greatly according to different authors.

Oskar Hertwig, for example, wrote in his book on the cell (1892), "The hypothetical idioblasts are small material particles in which the hereditary substance, or idioplasm, is decomposed. They are, according to the diversity of their material nature, the bearers of the morphological and physiological characteristics that we observe in the animate world. I will use two metaphors. First, idioblasts are comparable to the letters of the alphabet, which, though few in number, can in different combinations form different words, which can in turn be combined differently to form sentences with different meanings. Idioblasts are also comparable to the notes that engender such varied harmonies when they are combined or ordered in countless ways." We would not speak very differently today of our cistrons and codons.

In short, it can be said that by 1900 several theorists, meditating on the data of cytology and embryology, had formulated views fairly close to the concepts of modern genetics. But from an experimental standpoint, either they had no facts to offer in support of their conclusions, or they were unable to recognize and interpret those they had in such a way as to bring out their demonstrative value.

I have mentioned the experiments of Johann Dzierzon with bees, of Goss and Seton with pea plants; these were almost Mendelian experiments, and yet no one, so far as I know at least, tried to explain their remarkable results in terms of the chromosome theory of heredity. Only Weismann, perhaps, in a few passages of his works, sketched a timid connecton between his speculative views and certain data of observation or experiment.

THE MENDELIAN REVOLUTION

It was in 1900 that everything changed, with the Mendelian bombshell. At the beginning of that year a Dutch botanist, Hugo De Vries, published two papers on the hybridization of various plants. In one of them, published in Germany, he said that the essence of what he had discovered had been stated long before by a monk named Mendel, but in an article so seldom mentioned that he had not even known of its existence until after he had nearly finished his own work.

In April of that same year, a German botanist, Karl Correns, also announced results that were quite similar to those of Mendel. He, too, had at first thought that he was an innovator. Finally, in June 1900, an Austrian botanist, Erich Tschermak, corroborated Mendel's experiments. He had not learned of them until after he had finished his own. Who was this Mendel, whose work had just been triply rediscovered?

Johann Mendel's life was simple and obscure. He was born of peasant parents at Heinzendorf, Moravia, on July 22, 1822 (also the year of Louis Pasteur's birth). At the age of ten he entered the school at Leipnik; then, after having continued his studies at Troppau and the University of Vienna, he decided to become a monk. In 1843, he was admitted to the Augustinian monastery at Brünn as a novice, and in 1847, he was made a priest. His religious name was Brother Gregor.

Although he had no university degree, Mendel taught natural science and elementary physics at the Modern School of Brünn. He twice failed to pass examinations that would have qualified him for a higher position. He became an abbot in 1868 and died in 1884.

It was in 1856, at the age of thirty-four, that he began experimenting with hybrid peas in the garden of the monastery. His goal was at first quite modest: he intended only to produce new colors, for purely esthetic reasons, by artificial pollination. But as he continued crossing different species and extending the range of his work, his ambition grew; he realized that he had been led into the whole problem of heredity and that he needed to clarify it in order to interpret the results he was obtaining. He was struck by the fact that these results were so regular and clear cut that they could be expressed mathematically. He formulated hypotheses and performed new experiments in the hope of verifying them.

Finally, after having carried out thousands of artificial pollinations and examined tens of thousands of seeds, he felt that he was ready to state his conclusions in the form of general laws. He did so in a report entitled *Experiments in Plant Hybridization*, which he presented before the Brünn Society for the Study of Natural Science.

This report, which prompted two speeches delivered a month apart (February 8 and March 8, 1865) was nothing less than one of the most astonishing masterpieces ever to come from a human brain. In its fifty or so pages, describing the results of eight years of patient research, a whole science was revealed, and, still better, a new way of thinking in biology.

As Jacques Picquemal rightly said in a remarkable lecture (*Aspects de la pensée de Mendel*, Palais de la Découverte, 1965), "Mendelism is not merely a theory, however profound and extensible; it is a system of concepts and principles that created a new scientific field by showing the unity and relative autonomy of a certain domain, by defining in advance, in terms of its concepts and principles, the type of an unlimited number of problems to be investigated and the type of method to be used in seeking solutions to them."

In terms of importance and later consequences, Mendel's work can only be compared with that of Louis Pasteur on butyric fermentation, which appeared at about the same time. Everything that now constitutes the essence of the science of heredity, and everything in it that continues to develop in many different directions, was already contained, explicitly or implicitly, in Mendel's memorable report. Not one line of it has become obsolete, because it describes only flawless experiments and puts forward only hypotheses whose validity has since been confirmed.

Unfortunately, the very greatness and extraordinary newness of Mendel's work made it incomprehensible to his contemporaries. Despite his efforts to call it to the attention of official specialists—notably Nägeli, who had constructed an ingenious theory of heredity—he received no encouragement from them and finally abandoned his research.

Thirty-five years went by before his report was drawn from oblivion, and it might never have become known if Wilhelm Focke had not mentioned it briefly in a book on hybrid plants. The exhumation of Mendelism created a sensation in the scientific world.

The three scientists who, independently of each other, had almost simultaneously rediscovered what had already been discovered long before, unknown to them, by an obscure monk; that amateur botanist who, by the power of his genius, had outstripped the greatest authorities of his time; that marvelous report, buried in the periodical of a small local scientific society—all this caused astonishment and deep emotion.

As soon as it was rediscovered, Mendelism stimulated new research. Confirmation came in from all sides. The laws of hybridization—immediately named Mendel's Laws—were extended to animals by William Bateson in England and Lucien Cuénot in France. What was the exceptional significance of the Mendelian revelation and revolution?

First of all, for his experiments in hybridization, Mendel had chosen very stable varieties of peas with sharply defined differential traits: difference in the form (wrinkled or smooth) and color (green or yellow) of the ripe seed, in the shape of the ripe pod, in the color of the pod, in the length of the stalk. Furthermore, he had the brilliant intuition that, in order to find his way through the labyrinth of heredity, he would have to concern himself, not with the overall resemblance between parent and offspring, but with the presence or absence of a certain specific trait. Rather than thinking in terms of organisms, he thought in terms of traits.

From the results of his crosses he established laws that enabled him to predict the results of other crosses. He could predict, for example, that if he crossed two pea plants differing by certain traits—A in one, a in the other—all plants of the first generation would have trait A, known as dominant. He could also predict that when these hybrid plants were crossed, they would produce plants of type A and plants of type a, the former being, on the average, three times as numerous as the latter.

The predictions permitted by Mendelian analysis are, of course, statistical in nature; they are valid only on condition that the number of cases is high enough to allow the laws of probability to become significant. But who can fail to see the immense progress that Mendel had made? For the first time in the history of heredity research, the possibility of *prediction* had appeared. The apparent capriciousness of heredity had been overcome. One could no longer say, with Delage, "Anything is possible, and nothing is certain." There was now genetic impossibility, and at least statistical certainty.

Some meticulous commentators, examining the figures reported by Mendel, have claimed that it is improbable, nearly impossible, that his experiments should have given results so close to the theoretically predictable proportions. In other words, the agreement between fact and theory is too perfect, the experimental results too precisely, rigorously Mendelian, to be completely plausible. Perhaps an overzealous assistant, knowing what Mendel hoped to establish, "touched up" the statistics a little.

According to Jacques Picquemal, there can scarcely be any doubt that "the real figures, whether consciously or not, whether for didactic reasons or not, were altered to fit the previously formulated theory." This remark naturally does not affect the value of Mendel's experiments and concepts.

Mendelism enabled the investigator to predict the results of a given cross. This, of course, was extremely important. But furthermore—and this is where the Mendelian revolution was to have incalculably great consequences—the numerical relations discovered by Mendel were profoundly significant in that they revealed the working of essential and invisible mechanisms.

To account for the consistency of his results, Mendel was forced to make a certain number of hypotheses: he had to assume that each differential trait in his peas was related to an "element" that determined it and was transmitted by the reproductive cell; that these "elements," inherited from each parent and associated in the hybrid, were separated—divorced —in the hybrid's reproductive cells; and finally, that each couple of "elements" was divorced independently.

The idea of the *discontinuity* of the hereditary material, backed by experimental evidence, was now established. That material appeared to be formed of separable elements, basically analogous to the atoms of chemistry.

Mendelism thus gave support to the particular, "micromeristic" concept of heredity, to which many biologists had been led by pure speculation, as we have seen, by the end of the nineteenth century. It was therefore quite natural that the rediscovery of Mendelism brought about a connection between theory and experimental investigation.

Called *factors* before being named *genes* by Johannsen in 1903, the Mendelian "elements" were the subject of extensive research, which established the universality of Mendel's Laws and revealed how these factors act, the way in which they cooperate in the production of hereditary traits. A whole "factorial mechanism," often highly complex, was thus revealed. Little by little, it was to dispel the darkness that had so long enshrouded organic heredity.

It was at first thought that Mendel's Laws applied only to superficial, "ornamental" traits. (Félix Le Dantec compared Mendelian traits to a clown's many vests, which, when they have been taken off, still leave a whole man.) But the discovery of lethal factors (by Lucien Cuénot in 1905) showed that Mendelian factors could have basic physiological effects. In 1909 Sir Archibald Garrod discovered that a metabolic disorder which causes blackness of urine (alkaptonuria) is transmitted in the Mendelian manner; then von Dungern and Ludvig Hirszfeld showed that the inheritance of blood type obeys Mendel's Laws.

MENDELISM AND CHROMOSOMES

When the idea of factors had been established, questions arose. What are Mendelian factors, and where are they situated in the cell? Answers were soon given, because there was a striking similarity of behavior between these factors and the chromosomes in the cell nuclei of all living creatures. Some biologists, as we have seen, had already identified chromosomes with the hypothetical hereditary material.

It was quickly realized that the segregation of Mendelian traits corresponds to chromosome reduction in the reproductive cells, and the independent segregation of traits to the independence of the chromosomes. In 1902 William A. Sutton, a student of E. B. Watson, demonstrated this correspondence.

In the same period, Boveri's experiments with the egg of the sea urchin revealed the individuality of the chromosomes, their genetic continuity, and the specificity of their function in development, while McClung and Stevens showed the special function of a certain chromosome in determining sex in insects.

Then, beginning in 1910, a new and decisive step was taken: the admirable research of Thomas Hunt Morgan and his famous team (Calvin B. Bridges, A. H. Sturtevant, Herman Joseph Muller) on the fruit fly (*Drosophila melanogaster*). Because of its short life cycle, the ease with which it can be raised, the small number of its chromosomes, and other favorable features, the fruit fly was the ideal subject for research of the kind that Morgan undertook. It supported the chromosome theory of heredity with direct evidence, based on careful analysis of the many crosses that were made between the ordinary wild form of the insect and the numerous mutants observed in it, or between the mutants themselves. In the course of their laborious investigations, experimenters with fruit flies also used certain chromosome abnormalities (extra or missing chromosomes) which appeared in some lineages and caused the very irregularities in the transmission of traits that were predicted by the theory. This theory was repeatedly verified by observation of its consequences.

THE CHROMOSOME THEORY OF HEREDITY

The fruit fly has four pairs of chromosomes in its cells (one pair in the form of a rod, two V-shaped pairs, and one pair in the form of dots). If the genes are situated in the chromosomes, those that belong to the same chromosome should remain together in their transmission. And the fact is that certain genes are linked, that is, they form a group, and there are precisely four groups of genes, obviously corresponding to the four chromosomes.

Moreover, the genes belonging to one of these groups exhibit a special mode of heredity, sex-linked heredity, and one of the pairs of chromosomes differs according to sex (two straight chromosomes in the female, one straight and one bent in the male).

When experiments led to the conclusion that, during chromosome reduction, exchanges of genes (the process known as "crossing over") took place between corresponding chromosomes, the hypothesis of "chiasmatypy" (Janssens) was used, and it was assumed that two genes in the same chromosome separated more frequently in proportion to their distance from each other. It was thus possible to determine approximately the positions of the genes along the chromosomes.

By 1913, a chromosome map of the fruit fly (largely the work of Sturtevant) had been drawn up in its general outlines. It was corrected, made more precise, and completed after the discovery of the giant chromosomes present in the cells of the salivary glands of the larva, to which Theophilus Painter (1933) devoted detailed studies.

In 1925, Sturtevant reported the *position effect*, that is, he showed that the action of a gene depends on its location on the chromosome; and in 1927, Muller discovered that the frequency of mutations can be greatly increased by subjecting fruit flies to ionizing radiation.

This was a fundamental discovery, not only because it gave investigators a means of producing mutations at will but also because it led to better understanding of the nature of mutations and revealed the danger that can be presented to human heredity by short-wave radiation. Doctors were slow to take account of this danger, but in our age of "atomic peril" it has assumed major importance.

Almost simultaneously with Muller's work on fruit flies, Stadler reached similar conclusions with regard to plants. Fifteen years later, Charlotte Auerback produced artificial mutations in fruit flies by the use of chemical substances, such as mustard gas.

Parallel with research on genetics was the development of "physiological genetics," or the study of the processes by which genes produce their effects. Particularly important in this area was the research of Boris Ephrussi and George W. Beadle on the way in which "vermilion" and "cinnabar" genes modify the pigmentation of the eye buds in fruit flies by controlling the production of diffusible substances.

It would be impossible, in this rapid historical survey, to describe in detail how the idea of the gene was gradually developed. That idea has necessarily been altered since its origin, but perhaps not as much as is commonly assumed, because in the early days of Morganian genetics—and even before, in the speculative period—biologists were already speaking of the hereditary particles as if they were chemical molecules. Morgan hesitated to identify the gene as a molecule, but it seems clear that he was restrained only by scientific caution.

In any case, it is undeniable that in the Morganian period a sharp distinction was recognized between the gene and the "representative particle" of Weismann's theory. I therefore feel that Cyril Darlington is not quite fair to Morgan when he reproaches him (and excuses him at the same time) for not having had a sufficiently concrete idea of the gene. According to Darlington, Morgan's genes were as empty as Mendel's "elements" and Weismann's "ids." But I do not see how

Morgan, in his time, could have conceived the gene more precisely than he did. And he never believed his concept was final; he always assumed that the idea of the gene would be further developed in the future.

While the chromosome theory was being successfully studied in the fruit fly, other experimental subjects enabled genetics to expand in new directions. Albert Francis Blakeslee studied polyploidy and heteroploidy in a plant, the datura, in which he revealed the existence of as many mutations by addition of an extra chromosome as there are chromosomes in the genome. In a thorough study of a common mold (*Neurospora*), Beadle and his associates threw light on the biochemical activity of the genes by showing that each gene controls a series of chemical reactions which can be traced to the action of a specific enzyme.

After molds, it was bacteria that became a favorite experimental subject for geneticists, first because their speed of multiplication makes it easy to study mutations in them, and also because they provided the first examples—and so far the only ones—of certain mutations of a special type known as induced mutations. The discovery of these mutations resulted from a simple experiment. Although its importance was not immediately realized, it marked a new turn in the evolution of genetics.

DNA AND "HYBRIDIZATION BY CORPSES"

In 1928, the English bacteriologist Fred Griffith injected a mouse with living nonvirulent pneumococci and virulent pneumococci that had been killed by heat. To his surprise, the mouse died, and in its blood he found living virulent pneumococci.

The experiment was repeated, each time with the same result. The only explanation was that dead microbes were somehow able to transfer some of their traits—in this case, virulence—to living microbes. It was, as Alfred Mirsky has called it, "hybridization by corpses," an entirely new and unexpected fact. The transferred traits persisted in the descendants of the original microbes. There had been a hereditary variation: a mutation.

It was later demonstrated that the singular phenomenon could be reproduced, not only in the body of a mouse but also in a test tube, by mingling dead and living microbes. Then it was discovered that the dead microbes themselves could be replaced with extracts taken from them. Finally, in 1944, Oswald T. Avery and his colleagues Colin MacLeod and Maclyn McCarthy induced the mutation by subjecting living microbes to the action of a purified substance: deoxyribonucleic acid, or DNA.

From then on, DNA, capable of causing specific variations in the hereditary material, was naturally seen as the substance responsible for heredity. What is it?

It is a variety of nucleic acid. Nucleic acid was discovered in 1871 by a young Swiss chemist, Friedrich Miescher, who found it in pus cells. Later, Albrecht Kossel distinguished two kinds of nucleic acid: ribonucleic (in yeast) and deoxyribonucleic (in calf thymus glands), differing from each other in the nature of the sugar they contain. Delicate coloring techniques (Robert Feulgen) showed that all the DNA of a cell belongs to the nucleus, and more precisely, to the chromosomes.

Aside from the facts revealed by Avery, everything pointed to the fundamental importance of DNA in the transmission of hereditary traits. It is found in approximately constant quantity in cells and is always in proportion to their number of chromosomes (C. and R. Vandrely).

Concerning the genetic function of the nucleic acids, we

can mention a curious passage in Jacques Loeb's *The Dynamics of Living Matter* (1906), in which he wrote that in order to judge whether nucleic acid or protamine was the important factor in heredity, one would have to know whether the nucleus of the ovum also contained protamine, but that apparently it did not. He added that there seemed to be many more isomeric varieties of nucleic acid than of protamine or histone.

After Avery's experiment we leave the history of genetics, properly speaking, and enter its modern period. I will mention only that in 1953, James D. Watson and Francis H. C. Crick presented a model of the molecular structure of DNA: a structure composed of two strands forming a double helix. Each strand is a chain of nucleotides, and each nucleotide consists of a purine or pyrimidine base, a molecule of deoxyribose sugar, and a phosphate group.

In establishing this model, Watson and Crick used both chemical data (Erwin Chargaff had shown in 1951 that the respective frequencies of the different bases follow certain laws: in particular, the adenine/thymine and guanine/cytosine ratios are always close to one) and crystallographic data (X-ray diffraction pictures obtained in 1953 by Maurice Wilkins and his associates).

Specialists in all countries now began trying to understand how the complex properties of the hereditary material could result from different arrangements of a few relatively simple constitutents. They already foresaw the deciphering of the "genetic code" inscribed in DNA; they began to discover how nuclear DNA intervenes in the functioning of the cell and, by means of messenger RNA, controls the synthesis of proteins.

The study of heredity has now become essentially a matter of molecular biochemistry. We are far away from Réaumur's poultry, Mendel's peas, Morgan's fruit flies. And old-style biologists, biologists who are only biologists, feel a little disconcerted and perplexed by that new form of genetics, which continues to move farther away from them, and the respect they feel for it is mingled with a little melancholy.

Another surprise lay in store for geneticists: the swift development of human chromosome genetics, beginning in 1956. Before that time, very little was known about human chromosomes. It was generally accepted that there were forty-eight of them (twenty-four pairs), but biologists believed it would be a long time before much more was learned about them, for several reasons: the difficulty of obtaining cellular material in good condition, the large number and small size of the chromosomes, which made their enumeration difficult and uncertain. Scientists successfully studied and scrutinized the chromosomes of fruit flies, corn, daturas, irises, etc., but our human chromosomes, the ones that most aroused our curiosity, the ones we had the greatest practical interest in understanding, continued to elude precise analysis and remained almost inaccessible to research.

Then a new technique of examination—or rather an effective improvement on already existing techniques—was devised by Joe-Hin Tjio and Johan Levan, who produced cultures from the lungs of a human fetus. It now became possible, even relatively easy, to count the chromosomes in human cells and to differentiate the pairs of chromosomes, or at least most of them. Not only was it found that there are twenty-three of these pairs, not twenty-four, but anomalies in the number of chromosomes were discovered in abnormal individuals. Raymond Turpin, Jérôme Lejeune, and Gautier ascertained that mongoloids have an extra chromosome (a total of forty-seven, by trisomy of chromosome 21). Then there were successive discoveries of the addition of a sex chromosome in Klinefelter's syndrome, characterized by the karyotype XXY, the suppression of a sex chromosome in Turner's syndrome (karyotype XO), a slight chromosomal deficiency in the "cri-du-chat" syndrome,* and in certain types of leukemia (Philadelphia chromosome), etc.

In 1960, Lejeune and Turpin noted that discoveries were occurring at the rate of one or two a month, and the rate has not diminished since then; quite the contrary.

A whole chromosomal pathology—recalling that of the fruit fly—was rapidly established. It showed great promise, not only for medicine but also for anthropology and human biology. Thanks to it, new and unexpected concepts have already emerged—notably that of mosaic individuals, or the coexistence within a single individual of several cell populations, differing in their karyotypes, and that of monozygotic twins with different karyotypes (corresponding to a disjunction mosaic), and so on.

Here again we approach the swiftly evolving science of today. I simply want to note how surprising it is for a man of my age, who lived through the period of resignation, even defeatism, with regard to human chromosomes, when he learns that now, on hearing the tone of a newborn baby's crying, a doctor can diagnose the destruction of an arm of a certain chromosome.

The extent to which recent developments in human cytogenetics were unpredictable only fifteen years ago is illustrated by a personal experience that the eminent English geneticist E. B. Ford has related. One day, he and his associates were visited by an Oxford surgeon, Dr. Maloney, who was interested in the preparations of seminiferous tubules that they were making from mice. Maloney offered to supply them with human material. Ford thanked him, but thought to himself

---

* EDITORIAL NOTE: A chromosomal abnormality characterized by the timbre and plaintive tonality of the newborn infant's crying.

that although it might be interesting to look at human chromosomes some day when there was less work to do, it was already known that their diploid number was forty-eight and there was probably little more that could be learned about them for the time being.

The study of human genetics has now reached the stage that Morgan's school had reached by 1920. Doctors, previously outdistanced by biologists, have regained a certain advantage in this new direction taken by genetics and have particularly distinguished themselves.

LESSONS TAUGHT BY THE HISTORY OF GENETICS

In this brief and necessarily incomplete sketch of the history of genetics, I have had to pass over all the extensions of that science into cytology, serology, immunology, evolutionary genetics, etc. But even reduced to its simplest terms, the history of genetics teaches a number of lessons.

Let us examine the conclusions of genetics in relation to some of the basic problems of biology. Biology's past was marked by the great quarrel between preformationists and epigenesists. Developments in genetics have now shown us how hereditary traits are inscribed in the nucleic acids and can theoretically be read in them, provided we know the "code" of that molecular language. This does not amount to preformation in the classical sense, but it cannot be denied that there is a germinal preorganization and predetermination.

In the third edition (1925) of his famous book, *The Cell in Development and Heredity*, E. B. Wilson spoke of a "nuclear preformation," which, in the course of development, was expressed by a "process of cytoplasmic epigenesis."

There can be no doubt that men like Weismann and Nägeli would recognize our codons as descendants of their "biophores" and "micelles," which Delage and Le Dantec

derided, seeing in them "a disguised and shameful homunculus." And the great champion of total epigenesis, Paul
Wintrebert, attacked the embryologist Louis Gallien for saying
that allowances had to be made for a certain "germinal
preformation, represented by the genes."

It seems that, in this great debate, genetics has contributed
decisive arguments, if not in favor of genuine preformation, at
least in favor of a preorganization, a "preordination" rather
similar to what Charles Bonnet had in mind when he wrote,
with regard to the "germ": "This word designates not only an
organized body *reduced to small size* but also any kind of
original preformation from which a whole may result, as from
its immediate principle." (*Palingénésie philosophique.*)

It can also be said that the teachings of genetics have
permanently destroyed the tenacious belief in the transmissibility of acquired traits. It was already difficult to conceive
how a local alteration of the body could produce in the germ
cell the very mutation that would be expressed in the following generation by a somatic alteration of the same nature, but
who can now suppose that such a transmission could take
place by means of DNA? Lamarckism has truly been given
the *coup de grâce.*

There are a few more observations to be drawn from the
history of genetics. In this brief summary I have noted the
equally fruitful contributions of experiment, theory, and
technique.

The importance of well-conceived and executed experiments has been constantly apparent in the work of Mendel,
Morgan, Muller, Griffin, Avery. But theory has been no less
important. The speculative concepts of heredity that were
developed in the late nineteenth century remained completely sterile until Mendel provided a means of connecting
them with the facts, but the full significance of those facts

was appreciated only because they were clarified by hypotheses and set within the framework of a general theory.

The importance of technique is illustrated by the work of Tjio and Levan, which gave new impetus to the whole science of human chromosome genetics.

We have also seen the importance of introducing new material into a line of research that has exhausted its possibilities, or appears to have done so.

Scientific genetics was born of crosses between different species of peas, because peas lent themselves particularly well to the study of hybridization; but all of modern genetics could not have been built on peas. Use of fruit flies was decisive, because they were marvelously well suited to verification of the chromosome theory. Later, great advances were made by studying molds, then microbes, and finally man. Botany, zoology, bacteriology, and medicine have worked in close cooperation.

We must also note the part that physics and chemistry have played, and will continue to play, in increasing our genetic knowledge. The work of Crick and Watson would not have been possible without the contribution of chemists and even crystallographers.

Finally, we must note that there have been few fortuitous discoveries in the history of genetics. Nearly all advances have occurred regularly, logically, harmoniously, without fits and starts, and without any significant help from chance.

Like all scientific progress, progress in genetics has benefited from the widely varied qualities of the men who have contributed to it: the patience of Morgan, who bred fruit flies for years before he observed the first mutations; the visual acuity of Calvin Bridges, who distinguished a subtle variation in the eye of a fly, where his colleagues saw nothing; the intuition of Fred Griffith, who, faced with such an unexpected result as the transfer of a hereditary trait from a dead microbe to a living one, did not conclude that there must have been an experimental error; the ingenuity of those who worked out the structure of the DNA molecule; and all the other "factors of discovery," in which character and emotional reactions are no less important than intelligence and logic.

The development of genetics has not taken place without conflict, without stirring up resistance of various kinds, sometimes very vehement; and here, too, there are lessons to be learned.

Darlington has described his dismay on recalling the stubborn refusal of European biologists to accept the idea of the determinant, that is, the gene. Their fight against it, he says, was as energetic as the fight against the discoveries of mechanics in the seventeenth century, the phlogiston theory in the eighteenth, and the existence of animalcules, or microorganisms, in both centuries. He reports that a biologist once told him that such a simplistic idea could never have been born in Europe.

For my part, I remember—and such memories are instructive—the fierce opposition that Morganian genetics encountered in France. I have already spoken of that magnificant edifice of ideas, one of the most glorious acquisitions of modern science. While it was being built in America, small minds in France, infatuated with their false erudition, were rejecting the clear evidence of experiment in the name of Cartesian logic.

In 1937, Etienne Rabaud, professor at the Faculté des Sciences, in his book, *La Matière vivante et l'hérédité*, took the liberty of making ironic remarks about Morgan's "American simplicity" and the "singularly disturbing mentality" of that naïve man who had ignored difficulties, made arbitrary, contradictory, and often ridiculous statements, and used detestable, archaic, parasitic, and grotesque hypotheses to concoct a system that has proved worthless.

"How are we to describe, from a scientific point of view," wrote Rabaud, "a man who builds a system on such a shaky foundation and makes wild claims with so little justification?" An artisan of regression, a theorist who worked in a vacuum and whose influence was debilitating and discouraging, a maker of hypotheses that were only acts of faith and deserved no consideration in the country of Descartes and Claude Bernard, a man with a simplistic, backward mind who used deceptive language, contented himself with a sterile clatter of words, and relied on methods whose general application would amount to scientific suicide. Yes, that is how, only thirty years ago, a Sorbonne professor judged one of the creators of the science of heredity!

I have evoked that period—which is so painful for French biology and made us fall several decades behind in genetics—because I personally knew it, lived it, was subjected to it, when I began attending the Sorbonne more than half a century ago. I still remember the pitying look on the face of one of Rabaud's disciples when I confessed to him that the work of the Morganians did not seem to be entirely lacking in interest. Like everyone my age, I was deceived by the professors of that time. It is always good to point out, especially to young people, how far the blindness and smugness of certain pundits can go.

Now that genetics has made a fecund alliance with biochemistry, it is interesting to note that the adversaries of the chromosome theory—particularly Etienne Rabaud—demanded a *chemical* interpretation of the facts of heredity. According to them, a choice had to be made between morphology and chemistry. To believe in chromosomes, those "figurative elements" that had no more right to the biologist's attention than any other constituent of the cell, was to think as a morphologist and a vitalist.

But what do we see today? Contrary to the expectations of the anti-Morganians, the chromosome theory has led to the biochemical study of heredity and correctly guided the efforts of chemists by showing the essential function of the DNA contained in the chromosomes. To develop a valid and fruitful chemical approach, we had to pass through the morphological phase, the phase of the chromosome. And that, I believe, is an important lesson. Chemistry, of course, always has the last word; but in biology, at least, it must not make its demands too soon.

It is also interesting to note how many nonscientific—I will even say emotional—arguments were used against the chromosome theory. The mechanism of heredity could not be so simple, so mathematical; life was much more complex than that. Elements that could be colored and seen in the cell were no more important than those that could not. There was no noble, privileged substance in the cell; all its components had the same rank, the same functional dignity.

Finally, it is impossible not to recall, briefly, the relentless ideological offensive that was launched against Mendelo-Morganian genetics about 1948 by Soviet biologists of the Michurinian school, grouped around Lysenko.

After a long and tumultuous debate that took place before the scientific societies of the U.S.S.R., the Central Committee of the Communist Party passed a motion condemning Mendelo-Morganism. It was officially decreed that there was no hereditary material, that the gene was "as much a myth as the vital force," that Mendel's Laws—"the pea laws"—were false, that the Morganians were only "fly breeders," and that supporters of orthodox genetics were vicious reactionaries, fascists, obscurantists, idealists, lackeys of the militaristic bourgeoisie, clericalists, perverters, poisoners of the mind, enemies of the people, and so on and so forth.

Poor chromosomes: they were now "politicized," pressed into military service, and treated as dangerous adversaries of dialectical materialism! This time they were repudiated not in the name of Descartes, but of Marx!

The dictatorship of Stalin and Lysenko was a sad period for biology, but we must remember it as a deplorable example of what happens when political prejudice interferes with science. Amazing things were proclaimed by people whose ignorance was equaled only by their passion. Life was engendered in an egg yolk, grains of rye were formed on wheat stalks, cultivated oats produced wild oats, useful plants gave birth to weeds, etc.

In France, "Lysenkoism" was glorified by the poet Louis Aragon. He made himself an impromptu biologist in order to demolish Mendel's Laws (after all, they were invented by a monk); and the philosopher Garaudy confounded the leguminous reactionaries by revealing the insignificance of results obtained with "beans."

Today, things have returned to normal, at least in appearance. Lysenko has fallen from favor; but Soviet genetics has lost many years—as French genetics did earlier. And whenever science regresses or stands still in one country, the whole world loses.

What a lesson for free minds—*all* free minds! For it is certain that in this case the guilty party was not communism, but, much more generally, political or philosophical *sectarianism*, ideological fanaticism in all its forms.

Yesterday, that fanaticism was tinged with red; tomorrow, perhaps, it will have another color, and scientific truth, always the child of freedom of thought, will be rejected in the name of another delirium.

The hundredth anniversary of the publication of Claude Bernard's *Introduction à l'étude de la médecine expérimentale* has just been rightly commemorated. It is from that immortal book, a canon of scientific integrity, a Bible for those who want no Bible, that I will borrow my concluding words: "The role of the physiologist, like that of any other scientist, is to seek truth for itself, without trying to make it bear witness to the validity of a certain philosophical system. When a scientist pursues his research on the basis of a philosophical system, either he wanders into regions too far from reality, or the system gives his mind a kind of deceptive confidence and a rigidity incompatible with the freedom and flexibility that the experimenter must always maintain in his work."

HISTORY OF IDEAS ON THE ORIGIN OF LIFE

We are now used to the idea that every living organism comes from one or two parents, in other words, that life can arise only from already existing life. But that idea is a relatively recent acquisition of scientific thought. All through antiquity it was believed that life could be formed directly from elements of matter, and reputable scientists continued to believe it until the latter part of the nineteenth century.

We must recognize that this was the simplest, most natural view, the one that first came to mind. Like belief in the flatness and immobility of the earth, it seemed to be obviously supported by common sense, for we often see living organisms appear without having seen them preceded by others similar to them.

As proof that this opinion naturally comes to mind, we can cite the fact that even today, in nations with an advanced civilization, some people are still convinced that maggots are produced by rotten meat and earthworms by manure, that aphids are born from untended rose bushes, and that lice are formed in excessively long hair.

In this historical survey I will go back no further than Aristotle (384–322 B.C.), the first of the great naturalists. Aristotle made a distinction between animals that came from parents similar to them and those that had no parents, such as eels, which arose from the silt of rivers; fish, from marshes that had dried out and then become wet again; lice, from flesh; caterpillars, from plants; insects, from dew; and intestinal worms, produced by the transformation of excrement.

According to Diodorus Siculus (late first century B.C.), the sun fecundated the silt of the Nile; animals thus produced were often incomplete: their trunks could be seen struggling, while their hindquarters merged into the mud. Avicenna (980–1037) went so far as to suppose that new races of men were engendered by human corpses washed ashore. In 1542, Jerome Cardan wrote that water engendered fish and that rats were born of putrefaction. All this is rather monotonous.

It is important to know the opinion on spontaneous generation once held by the Church. It was a rather ambiguous opinion and was altered through the centuries. The Bible teaches the special creation of each living species, but in the Book of Judges it is said that a swarm of bees issued from the body of a dead lion. Hence the riddle that Samson put to the Philistines: "Out of the eater came forth meat, and out of the strong came forth sweetness."

Several Fathers of the Church accepted spontaneous generation, each in his own way. Saint Augustine (354–430) said that it was due to the persistence of divine powers in matter. All living creatures produced by matter were created potentially and materially on the fifth and sixth days. These "seminal principles" waited different lengths of time before manifesting themselves: whenever a living creature appeared spontaneously, it meant that one of the principles had encountered conditions favorable to its development.

Saint Thomas Aquinas (1226–1274) took a different view of spontaneous generation. For him, there were no "seminal principles." The spontaneous birth of a living creature was a genuine beginning; life appeared only at the moment when an organism capable of receiving it had been constituted.

Let us leave the subtleties of theology and move on to the sixteenth century.

Jan Baptista Van Helmont (1577–1644) gave precise instructions for causing fleas and other animals to arise from "ferments" contained in matter.

First experiment: Pour pure fountain water into a container impregnated with the odor of ferments; molds, worms, and mosquitoes will be produced.

Second experiment: Marsh ferments engender mollusks, snails, leeches, and grasses.

Third experiment: Crushed basil leaves give off a ferment that generates scorpions.

Van Helmont's most famous experiment is the following. A woman's chemise, worn until it becomes dirty, gives off a ferment which, acting on grains of wheat, transforms them into adult mice of both sexes; twenty-one days are required for the transformation.

With Father Athanasius Kircher (1602–1680), we find a new way of conceiving spontaneous generation. According to Kircher, all parts of an animal are filled with small, highly volatile bodies: animal or seminal spirits. When an animal dies, they escape from the corpse, join a certain quantity of fixed matter, and form animals that are thoroughly alive, but smaller and less noble than the one from which they came. The "spirits" liberated by the death of a horse, for example, may be sufficient to form a fly, or even a frog. The process imagined by Kircher is thus midway between spontaneous generation, strictly speaking, and ordinary generation.

Kircher also claims to have seen thousands of ants arise from the body of a single dead ant. If the roads are full of earthworms after a shower, it is because the rain has moistened the dried bodies of dead worms. Hence, by analogy, a good recipe for making snakes: take as many dead snakes as you like, dry them out, cut them into little pieces, bury them in damp earth, sprinkle the earth abundantly with rainwater, and then let the spring sunshine do its work. A week later, small earthworms will appear. Fed on milk and earth, they will become so many perfect snakes, capable of propagating themselves by ordinary generation.

In the eighteenth century, the great entomologist Réaumur wrote in his memoirs that he had tried some of the experiments described by Kircher: "I admit with a kind of shame that I sowed earthworm powder, with the precautions indicated by Kircher, and buried pieces of dried worms like plant cuttings, without ever causing a single earthworm to appear. I had to have a perfect right to say that those claims are false, in order to give a satisfactory answer to people who feel that there can be no kind of evidence against what they believe to be true."

Kircher gave a long and minute description of Noah's Ark; and he noted, of course, that it was completely unnecessary to load the vessel down with animals that were capable of spontaneous generation. We shall see later that Noah's Ark has a certain importance in the history of spontaneous generation.

I must say a few words about the great physiologist William Harvey (1578–1657), because he is said to have opposed the idea of spontaneous generation. Actually, however, his famous axiom, "*Ex ovo omnia*," did not have in his mind the meaning that is usually attributed to it. For him, the egg (*ovum*) was simply a more or less egg-shaped *primordium*, which could be formed directly from inanimate matter. It was thus that intestinal worms, lice, and itch mites were born of the human body or its excrement.

Francis Bacon (1561–1626) also had no doubt of spontaneous generation, at least so far as many inferior plants and animals were concerned. And he maintained that study of spontaneous generation would be fruitful for physicians and philosophers. Since nature formed living creatures, man should also be able to do so, for in Bacon's general outlook, man could do anything that nature did, and even more.

Bacon therefore urged that efforts be made to determine experimentally the general conditions of the "natural vivification" that was spontaneous generation, and the specific conditions that suited various species: which animals were formed in living wood, or in dead wood, or in certain kinds of wood, or at various times of the year, and whether they needed rain, dew, shade, or sunlight. This experimental study of spontaneous generation would not only help to clarify its causes but also lead to the discovery of many facts concerning the nature and configuration of fully developed animals.

The English physician and writer Thomas Browne (1605–1682) held that frogs were born of rotting matter and lice from excremental humors. With regard to lice, he wondered why, since, after all, they come from us, we did not love them as our children. He did not believe in spontaneous generation for the larger animals: if death created life instead of destroying it, he said, and if graves were the most fertile wombs, then Noah's Ark would have been useless.

Sinibaldi denied spontaneous generation, arguing that if animals could be engendered in that way, God would not have ordered Noah to take couples of all living creatures into his Ark. And when Father Marin Mersenne conscientiously calculated the tonnage of the Ark, he made the proper correction by excluding all animals that could be born of corruption, such as lice, flies, and mice.

Let us now take stock of the situation in the middle of the seventeenth century, because we are approaching the fundamental date. At that time, spontaneous generation was accepted by everyone for the lower animals: worms, insects, snails, etc. Many accepted it for fish, and some even for mice. Thus Ross, objecting to Browne's skepticism, said that he might as well doubt that worms were formed in cheese, or snails, eels, and wasps in corruption. To dispute the spontaneous generation of mice, he added, was to oppose reason, common sense, and experience. Anyone who doubted it had only to go to Egypt, where he would see fields swarming with mice formed in the silt of the Nile, to the detriment of the human inhabitants.

It was in 1668 that the great blow was dealt to the ancient belief in spontaneous generation. That year is a memorable, essential date that everyone ought to know; it is much more important in human history than the dates of battles or treaties.

The physician, naturalist, letter-writer, and poet Francesco Redi (1626–1698) began by sharing the misconception of his time: like everyone else, he believed in spontaneous generation, at least for the lower animals. He even tried to reproduce Kircher's experiments in which plant stalks soaked in water were supposed to turn into animals.

But one day he reread a passage of the *Iliad* in which Achilles, holding the dead body of Patroclus in his arms, expresses fear that flying insects may penetrate his friend's wounded flesh and spread corruption in it, and his mother Thetis replies, "My son, I will shield him from the ardent swarm that consumes the victims of Mars."

This was a revelation to Redi. The "ardent swarm"? Did this not refer to flies, and was it not actually they that produced the maggots which supposedly came from rotting flesh? Redi's whole merit lay in asking himself the question. His experiment followed from it naturally.

In four wide-mouthed jars he placed, respectively, a snake, a few freshwater fish, four eels from the Arno River, and a slice of veal. Then he closed the jars by tying paper tightly over their mouths with string. He placed similar objects in a "control group" of four other jars but left the mouths open.

Flies went in and out of the open jars. Within a few days, worms were wriggling in the snake, the fish, the eels, and the veal that they contained. But not one worm appeared in the closed jars, even after several months.

Since it might have been objected that he had hampered the production of maggots by preventing air from coming into the closed jars, Redi improved his experiment. This time, instead of using paper to close four of the jars, he used cloth woven tightly enough to keep out flies but not so tightly that air could not enter freely. The result was exactly the same as before.

Redi had made an immense discovery, because it enabled him to conclude that maggots were not spontaneously generated in flesh and that they were hatched from eggs laid by flies. His experiment was so magnificently simple that it seems surprising that no one had ever thought of it before, but the idea of it could have come only from a free, independent mind capable of placing commonly accepted beliefs in doubt.

The book in which Redi described his results (*Esperienze intorno alla generazione degli insetti*) appeared in 1668 and was highly successful: seven editions in twenty years.

In 1687 the great microscopist Leeuwenhoek confirmed the experiment, using pieces of human flesh.

Redi's ideas were attacked at first, particularly by those who charged him with having contradicted the Bible. Here is a passage from his historic book, in which he states his credo of opposition to spontaneous generation: "Trying in this, as in all other things, to let myself be corrected by those wiser than I when I am mistaken, I will not deny that as the result of the many objections that were presented to me, I was inclined to believe that since the first days of the world, when the sovereign and all-powerful Creator commanded the earth to bring forth the first plants and animals, it has never produced of itself a single animal, tree, or plant of any kind, perfect or imperfect, and that everything which was born in or of it in the past, and is so born today, comes from the true and real seed of the same plants and animals, which preserve their species by means of that seed." (*Esperienze.*)

Redi felt, however, that evidence against spontaneous generation was insufficient in the case of larvae lodging in fruit, vegetables, leaves, and trees. To explain the presence of those larvae within plant tissues, he considered two hypotheses, without daring to choose between them: either the worms came from outside, seeking food, and gnawed their way through the tissues, thus reaching the interior of the fruit or tree, or else the worms were engendered by the same spirit and the same force that engendered flowers and fruit.

The problem of insects living in plants, particularly those insects that produce galls, was solved by Marcello Malpighi (1628–1694), a great anatomist and microscopist, author of the treatise *De Gallis et Plantarum Tumoribus et Excrescentiis.*

Malpighi observed small flies laying their eggs in the buds and leaves of trees; he concluded that a gall was a tumor formed around an egg by the growth of the plant tissues.

Having caught a fly just when it had inserted its ovipositor in an oak bud, he ascertained that its abdomen contained eggs exactly like those he had found in the bud.

Other facts of the same order were reported by Antonio Vallisnieri (1661–1730), an Italian physician and naturalist who studied with Malpighi in Bologna and practiced his profession in Scandiano from 1689 onward.

The Dutch naturalist Jan Swammerdam (1637–1680) was a great adversary of spontaneous generation. An astonishingly skillful dissector, he wondered at the complex organization of even the smallest insects. He examined lice, mayflies, mites, and bees with the passionate ardor of a scrupulous and religious anatomist who believed he saw the mark of divine creation in the perfection of organic detail that he discovered.

He felt it was madness to claim that those wonders, those masterpieces of structural delicacy, could arise fortuitously from corruption. It was "more the opinion of a beast than of a human being" to believe that those amazing animated machines, whose secrets could not be exhausted by long and painstaking scientific investigation, were produced by blind chance in the twinkling of an eye. For anyone able to see and reason, there was no difference between the large and the small, there were no animals more perfect or noble than others: if a louse could be born of sweat, then a man could be born of a dungheap.

It can be said, in general, that by the end of the seventeenth century the theory of spontaneous generation was out of favor with regard to insects, worms, and all other animals visible to the naked eye, although the contrary theory still encountered a number of difficulties, notably in the origin of intestinal worms. But then belief in spontaneous generation was strengthened by the discovery of microorganisms.

Under the influence of Swammerdam in particular, use of the magnifying glass and the microscope had contributed to the decline of "spontanism" by revealing the complexity of small animals. But finally the microscope gave "spontanists" an opportunity to carry on their fight in another area.

In 1674—another important date—the Dutch naturalist Anton van Leeuwenhoek (1632–1723), examining a drop of stagnant water under a microscope, was astounded to see that it was swarming with life.

Here is an extract from the letter he wrote from Delft to Mr. Oldenburg, Secretary of the Royal Society of London, on September 7, 1674:

"Two hours away from the village there is a lake, called the Berkelse Mere, whose bottom is very muddy in many places. Its water is quite clear in winter, but by the beginning or middle of summer it turns whitish, with little green clouds floating in it. . . . I gathered some of it in a glass bottle and, examining it, I saw particles of earth and green filaments in it. . . . It also contained a great number of small animalcules [microorganisms] of different shapes and colors. . . . And the movement of these animalcules in the water was so quick and varied that it was a marvelous spectacle. I believe that some of these tiny creatures were less than a thousandth as large as the smallest I had ever observed before, in cheese, flour, molds, and elsewhere. . . ."

Leeuwenhoek mentioned these "incredibly small" animalcules again in December 1675, and in 1676 he spoke of animalcules of the same sort that he had found in rainwater, infusions of pepper, and other infusions.

In 1677, referring to the difficulty of accepting the idea that a single drop of water could contain so many living creatures, he agreed that it was not easy "to understand such things if one has not seen them." To support his own testimony, he produced written statements signed by "gentlemen" of his acquaintance.

Trying to estimate the number of animalcules, he gave these figures: between 2,730,000 and 8,280,000 in a small drop of water, between 10,000 and 45,000 in the volume of a millet seed, and about 27,000,000 in the volume of a grain of sand. These observations caused a stir among scientists, and even among intellectuals of all kinds.

As for the origin of the animalcules, Leeuwenhoek's first thought was that they were formed by "fortuitous segregation of the particles of water," but he soon rejected that explanation in favor of the assumption that the animalcules, or their seed, preexisted in rainwater. He performed experiments in an effort to clear up the problem.

"When I had learned of the various opinions expressed with regard to the genesis of animalcules," he wrote in a letter dated June 14, 1680, "and when I had learned in particular that a certain gentleman had stated that no living creature could be produced from meat or a mold if it was kept in a carefully closed bottle or vase, I resolved to make tests of that kind." The "certain gentleman" was obviously Francesco Redi; not knowing Italian, Leeuwenhoek was acquainted with his work only by hearsay.

Leeuwenhoek collected some rainwater "in an old china plate that had not been used for ten years" (he often reported such picturesque details), placed it in two glass tubes, and sprinkled a little pepper into it. He then closed one of the tubes by melting its end over a flame, leaving the other tube open. Several days later, there were as many animalcules in the closed tube as in the open one. He concluded that while Redi's conclusions might be valid for insects, they did not hold for animalcules.

He did not pursue these experiments. They could not have given him a clear answer in any case, because they involved many sources of error that he could not even have suspected.

The problem of the origin of animalcules was not close to solution. Two factions remained in opposition, and it must be said that there were good arguments on both sides.

Spontaneous generation was undoubtedly an exception to the great law of the continuity of life. But how was the appearance of animalcules in an infusion to be explained? Was it possible to accept the idea that seeds—"germs"—pre-existed everywhere and at all times, either in the liquid itself, in the matter that was infused into it, or in the atmosphere? As often happens in science, a choice had to be made between two hypotheses, neither of which seemed entirely satisfactory.

About 1740, a vigorous offensive was launched by the spontanists, headed by the great Buffon himself and his associate John Needham, an Irish Jesuit priest. Needham claimed nothing less than to have seen eelworms being produced in flour made from ergotized wheat. This report delighted Buffon, because it supported his theory, according to which every living creature was formed of "organic molecules" capable of producing other living creatures of a lower order after death.

For Buffon, who was in essential agreement with Athanasius Kircher, there were perhaps as many creatures produced by decomposition or corruption as by ordinary generation. It was this process that produced earthworms, grasshoppers, caterpillars, lice, eelworms, and, of course, all the animalcules found in infusions. It was therefore unnecessary to study these animalcules separately, which meant that a great deal of time could be saved by those who, with their microscopes, thought they were making observations that deserved to be called discoveries.

The resurgence of such ideas in the middle of the eighteenth century represented an enormous regression. Although

Needham's eelworms were enthusiastically received by Buffon, Voltaire attacked the "eelworm Jesuit" with his sarcasm: "The first man who said that there is no foolishness of which the human mind is not capable was a great prophet. . . . An Irish Jesuit named Needham, traveling in Europe in secular clothes . . . looked at some flour made from ergotized wheat . . . and thought he saw eelworms that soon gave birth to other eelworms. . . . Several philosophers immediately began proclaiming a wonder and saying that there was no germ. . . . Good naturalists were deceived by a Jesuit. . . . It seemed unquestionable that since flour made from bad wheat formed eelworms, flour made from good wheat could produce men."

Voltaire was right in this case, but to be fair to the "eelworm Jesuit" we must point out that he was the author of an experiment which had far-reaching consequences.

Needham filled a bottle with mutton gravy, closed it with cotton, and held it for a certain time over hot embers, so as to destroy any germs that might be in it. Despite this heating, which was intended to assure what we would now call the sterility of the liquid, numerous animalcules appeared in it. To Needham, this proved that they were produced by spontaneous generation from the infused matter. His conclusion was wrong, of course, but the procedure, the technique that he had used for the first time—in 1745—proved extremely fruitful later.

All through the long debate on spontaneous generation, lasting until about 1870, Needham's experiment was repeated over and over again by men who modified its conditions, trying to improve it and eliminate all sources of error. We must do Needham justice on this point: we are indebted to him for the technique by which his own assertions were disproved.

Needham's experiment was vehemently criticized by the antispontanists. They raised all sorts of theoretical objections against it. "We have laughed at Epicurus," wrote Charles Bonnet, "for forming a world from atoms; is it any less contrary to sound philosophy to form an animal from mutton gravy?"

But the partisans of germs limited themselves to rejecting their adversaries' conclusions without exposing the weak point in their position. It was not until Spallanzani entered the debate that it was finally placed on the level of experiment, where it remained from then on.

About 1770, Lazzaro Spallanzani showed that Needham's experiment had two sources of error: first, the bottle was imperfectly sealed by cotton; second, the duration and degree of heat were insufficient to make sure that all germs had been destroyed. If the experiment was performed with precautions that prevented the introduction of germs from outside, the results were quite different: the liquid did not become murky and no animalcules appeared in it.

This still did not resolve the question, however, and Needham did not admit defeat. He argued and quibbled about the value of the Italian biologist's version of the experiment. He claimed that overheating the infused matter destroyed the "vegetative force" necessary for combining organic molecules into living creatures; he maintained that torturing nature forced her "to give false testimony." Spallanzani replied with new experiments that were more and more refined, more and more embarrassing to Needham and the other spontanists.

Even then, the debate was not settled; it could not have been at that time, when almost nothing was known about the conditions that microorganisms require in order to live, particularly their need for oxygen. But although Spallanzani's experiments allowed a few sources of error to remain, on the whole they were accurate and well conceived, as ingeniously devised and skillfully executed as they could have been with the knowledge and techniques at his disposal. They were headed in the right direction, and they tended to validate a hypothesis whose extreme fecundity would be demonstrated in the future.

The debate between spontanists and antispontanists had now taken the form it would keep for a long time. Whenever the antispontanists set up experimental conditions such that no animalcules appeared in the medium, the spontanists objected that, by trying to preclude the intervention of germs, they had disturbed the conditions that permitted life to be formed. And whenever the spontanists presented facts that seemed to testify in favor of spontaneous generation, the antispontanists objected that they had not taken all precautions necessary to assure the exclusion of germs.*

It was hard to see a way out of this impasse, because it was quite true that procedures capable of preventing the intrusion of germs were often of a nature to create artificial conditions that could theoretically be accused of forming an obstacle to the birth of life.

About 1800, Nicolas Appert, a former confectioner living on the rue des Lombards in Paris, cleverly perfected a method of food preservation that had been proposed long before. He popularized it in a book entitled *Livre de tous les ménages, l'art de conserver pendant plusieurs années toutes les substances animales et végétales.*

Appert placed fruits or vegetables in hermetically closed

---

* THERE was, notably, the problem of the germs' resistance to heat. When life appeared in a heated medium, the antispontanists maintained that the heat had not been sufficient to destroy all germs, to which the spontanists retorted, "What do you know about the temperature necesssary to destroy germs? You are in a vicious circle: you cannot conclude that all germs have been destroyed unless life does not appear after heating."

containers and then heated them for a long time in boiling water. After this simple treatment, the food could be kept for years in a state of perfect preservation and edibility.

"Monsieur Appert," commented an article in *Le Courrier de l'Europe*, February 10, 1809, "has found a way to make the seasons stand still. For him, spring, summer, and autumn live in bottles, like those delicate plants that a gardener places under glass domes to protect them from the inclemency of the seasons."

Basically, Appert's jars were Spallanzani's bottles adapted to the needs of household economy. To the antispontanists, they were proof that substances highly subject to putrefaction remained unputrefied when they had been sterilized by heat —that is, when they had been made free of germs.

In 1810, the great chemist Joseph Louis Gay-Lussac had the idea of analyzing the air contained in Appert's jars. Finding no oxygen in it, he concluded that absence of this element was a necessary condition for the preservation of organic substances. And this conclusion was enough to reanimate the whole debate on spontaneous generation.

As we have seen, Needham maintained that when Spallanzani applied heat for a long time to his bottles containing infusions, he altered their inner atmosphere in such a way as to prevent life from appearing in them. On the basis of Gay-Lussac's observation, was it not plausible to assume that these bottles consistently remained devoid of life only because the oxygen had been removed from them?

It was now important for the antispontanists to repeat Spallanzani's experiments, taking the new objections into account. They had to obtain the same result as Spallanzani, that is, permanent sterility of the infusion, but under conditions that would allow the infusion to remain in contact with air of normal composition, containing oxygen. Unless it could be demonstrated that life was incapable of appearing spontaneously, even in the presence of oxygen, the spontanists would be entitled to maintain their position.

It was a German, Frank Schulze, who performed the necessary experiment in 1836. (In those days science did not advance with long, rapid strides as it does today.) He placed an infusion in a tightly sealed balloon flask, boiled it, and then introduced a current of normal air that had previously been bubbled through sulfuric acid. The purpose of the acid, in Schulze's mind, was to destroy any germs that might be present in the air but without altering its oxygen content. The result was that the infusion did not become murky, and no life appeared in it. Schulze felt justified in concluding that life was not formed in the absence of germs. Antispontanism had scored a point.

It scored another the following year, when the biologist Theodor Schwann (one of the two founders of the cell theory) presented an experiment of the same type as Schulze's, but differing from it in that the air reached the infusion after having passed over red-hot metal, rather than through sulfuric acid. With an infusion consisting of meat broth (a yeast infusion gave highly inconsistent results), no life appeared—in most cases, at least.

Schwann cautiously concluded, without using the compromising word, "germ," that the birth of life required the presence of something that could be destroyed by heat, or thermolabile, as we would say today.

Finally, in 1854, Schroeder and Dusch performed a third experiment, in which, rather than making the air bubble through an acid or heating it to a high temperature, they simply filtered it through cotton. It was a remarkable idea, but the results were not clear enough to justify a categorical conclusion.

Did spontanism lay down its arms? By no means, and it must be acknowledged that for the moment the experimental results were not compelling. First of all, they were inconsistent: the experiments did not succeed in every case, with all culture mediums. And even the successful experiments could be interpreted without accepting the hypothesis of germs. It might be that heat, sulfuric acid, or cotton deprived normal air of an unknown "something" that was indispensable to spontaneous generation. It could therefore be said that, although antispontanism had made definite progress, it had not yet won victory. Its adversaries had retreated, but the positions to which they had withdrawn were still defensible.

Beginning in 1858, the spontanists attempted a counter-offensive, led by Félix-Archimède Pouchet, Director of the Rouen Museum of Natural History. He was an enthusiastic, passionate man, not entirely lacking in worth as a physiologist, naturalist, and scholar, but he was simply unqualified to approach a subject so delicate and filled with pitfalls as that of spontaneous generation.

Pouchet began with a simple experiment that seemed conclusive to him. He filled a bottle with boiling water, sealed it hermetically, turned it upside down, and placed its neck in a basin of mercury to preclude any possibility that germs might be brought into it by air. Then, when the water had cooled, he opened the bottle and introduced pure oxygen into it, along with some hay that had been charred by keeping it heated to a temperature of 100 degrees centigrade for thirty minutes. Within a few days the hay infusion was filled with living organisms, which, according to Pouchet, could have no other origin than spontaneous generation. He did not realize the multiple sources of error that vitiated his experiment.

He performed many others, some quite ingenious, but all

equally open to criticism. He developed his theory at great length in a thick and bombastically written book: *Hétéro-génie ou Traité de la Génération spontanée* (1859).

In order for life to appear, said Pouchet, three conditions were necessary and sufficient: a substance subject to putrefaction, water, and air. The putrescible substance could be boiled or charred; the water could be filtered, distilled, or prepared by chemical synthesis; the air could be heated or passed through an acid, and even artificial air could be used. But no matter what treatments were applied to the putrescible substance, the water, and the air, life appeared whenever they were brought together. What more could anyone ask?

Then everything was clarified and explained; the problem was solved at last, with as much finesse as genius, by a chemist: Louis Pasteur.

Born in 1822 at Dôle, on the rue des Tanneurs (his father was a tanner), Pasteur was brought up at Arbois. Early in life he showed an atonishing talent for drawing and pastels (he nearly became a painter, as Claude Bernard nearly became a playwright), but, attracted to science, he attended the École Normale Supérieure. At the age of twenty-five, he made a splendid discovery concerning the dissymmetry of crystalline forms. In 1857, he approached the problem of ferments. In opposition to chemical theory of the time, as expounded by men like Baron Jöns Jacob Berzelius and Baron Justus von Liebig, he showed that alcoholic and lactic fermentation was produced by organic agents, or yeasts. Investigating the origin of ferments, he was involuntarily led into the problem of spontaneous generation, because fermentations often occurred "spontaneously," in the sense that no yeasts had been deliberately introduced into the fermentable medium.

Where did those yeasts come from? Pasteur approached the

problem without preconceived ideas. In 1859, after having shown that lactic yeast could be cultivated in a medium with a specific composition, he acknowledged that the question of spontaneous generation was still "open and totally lacking in decisive evidence."

But in 1860—that is, two years after Pouchet's first publication—Pasteur took a firm stand in favor of antispontanism. From then on, the two men were in ardent conflict with each other. It was an uneven contest if ever there was one (in no sense a "battle of giants," as it has been described), but it had the advantage of repeatedly stimulating Pasteur's imagination.

In 1861, he presented a report to the Institut National de France on "small organized bodies that exist in the atmosphere." The very subject of his work showed that he had detected the presence of those germs about which had been so often discussed, but whose existence had not yet been proven.

Pasteur had passed air through guncotton and collected the matter filtered out of it. The "small organized bodies," or "germs," were recognizable in this dust when it was examined under a microscope. And if the dust was carefully placed in a sterile flask containing a putrescible medium, or even a medium that had a mineral composition but was capable of sustaining the life of ferments, the medium thus "seeded" became murky and was soon populated by microorganisms.

Pasteur reproduced the experiments of his predecessors, notably Schwann, but with increased precautions. They consistently gave results that tended to support the hypothesis of germs. And, more important, he devised and performed new experiments. His "swan-necked flask" experiment is justly famous.

He heated the neck of a glass flask and stretched it into a long, winding tube which he left open to the outer air. The flask contained a fermentable liquid. After being boiled, this liquid did not become murky, even though it remained in contact with normal, natural air that had not been heated, filtered, or chemically altered.

Why did it not become murky? Because when air rushed into the flask after the liquid had been boiled, it was sterilized by the heat of the tube, and afterward, slowly moving through the winding passage, it deposited its germs on the glass wall of the tube. If the neck of the flask was filed off, the liquid soon became murky.

In its extreme simplicity, the experiment was highly significant: *for the first time* a putrescible liquid, left in contact with perfectly natural air, could be seen remaining sterile indefinitely because the arrival of germs was intercepted.

Today, in the Palais de la Découverte, there are some of those flasks that have remained sterile for nearly a century, venerable relics that, long after Pasteur's death, still prove the truth of his assertions.

He also showed that germs were by no means uniformly distributed in the atmosphere: "Some portions of the air contain no germs." He filled several flasks with an infusion and boiled it; then, when air was admitted into all of them, microorganisms appeared in some but not in others.

When the experiment was performed on a mountaintop, the percentage of sterile flasks was greatly increased, because the atmosphere contains fewer germs at high altitudes. At Montenvers, on the Mer de Glace glacier, nineteen out of twenty flasks remained sterile.

Finally, in 1863, Pasteur performed an experiment that was perhaps even more conclusive than any of those that had preceded it. He gathered *fresh* organic liquid (blood or urine) and placed it in contact with sterile air; no microorganisms or ferments appeared.

Only then was he fully satisfied, because he had previously wondered, without saying so, whether heating the fermentable substance might not cause some subtle alteration capable of preventing the birth of life. In his own words, the final blow had been dealt to the doctrine of spontaneous generation.

Pouchet, of course, was not convinced by Pasteur's clear demonstrations. He contested the reality of the facts. Had anyone ever actually seen those famous germs in the atmosphere, which Pasteur had supposedly shown to everyone in Paris? "What would we think of a scientist who claimed that the whole atmosphere was full of poppy, hemp, or lentil seeds, but could never place a single one of them before anyone's eyes? We would turn away from him, smiling."

Pouchet had examined dust of all kinds and all ages, gathered in all sorts of places: no germs, except for an occasional "genuine and rare exception" among grains of starch and granules of silica. "Pasteur sows the invisible and reaps only what must arise spontaneously." What could be more ludicrous than that hypothesis of "panspermy"! It was an affront to common sense, a scandal to the mind. Carried to its logical conclusion, it would require the assumption that each cubic millimeter of air contained 6,250,000,000 germs: "The air in which we live would have the density of iron."

Furthermore, why should the Creator have scattered germs so profusely for the sole purpose of populating a few puddles of water? Was it not an insult to His supreme wisdom to attribute such prodigality to Him?

As for the hypothesis of "limited panspermy," it was nothing but a sad defeat. And it was not true that flasks remained sterile when they were opened at the top of a glacier. To prove Pasteur wrong on this point, Pouchet climbed one of the Maladetta Mountains.

It is instructive for a historian of science to reread Pouchet's *Hétérogénie*, in which false logic, false precision, and false rigor are glaringly obvious on every page. Thus, to prove that the species of microorganisms varied with the nature of the infused substance, Pouchet macerated fragments of human skulls from different nations and different historical periods; he then triumphantly announced that an infusion of ancient Egyptian skull fragments (from the necropolis at Saqqara) produced epistylises, enchelyids, and vibrios, while Merovingian skulls, taken from burial vaults dating from the time of Merovaeus and Chilperic, produced glaucomas and vorticellidae, and modern skulls produced only colpoda.

Pouchet carried scrupulousness to the point of using "artificial air," a mixture of oxygen and nitrogen; this experiment, he felt, "speaks with such authority that it seems impossible to offer a bolder denial to partisans of aerial panspermy."

One of his disciples, Musset, used air filtered through animal membranes (ceca) and, still better, air taken from the swim bladders of fish or the insides of pumpkins.

All these spontanists were intelligent, ingenious, and sincere, and many of them were even good naturalists; but they had no idea of how the problem could be solved. They did not even suspect the kind of precautions that would have been required in order to make their results valid. They were unaware that in every phase of their experiments they opened the door wide to the germs they claimed to exclude and that all their efforts to achieve scientific rigor were made futile by enormous mistakes in method.

Pasteur was far from having the unanimous support of other scientists in that heated controversy. The Italian physiologist Paolo Mantegazza wrote, "I have always believed in spontaneous generation, but, after having visited Pouchet in his laboratory, after having seen his experiments and noted the precision with which they were carried out, and after having seen the ingenious apparatus used by him, I now believe in spontaneous generation twice as firmly as before."

Moreover, philosophical preconceptions and even political bias entered into the quarrel and led to distorted judgments. Although he claimed never to mix science with metaphysics, Pasteur was imprudent enough to speak these rather provocative words during a lecture at the Sorbonne: "What a victory it would be for materialism, gentlemen, if it could be supported with incontrovertible evidence for the view that matter forms itself into living organisms! What need would there then be for the idea of a primordial Creation before whose mystery everyone must bow? What need would there be for the idea of a creating God?"

And Father Moigno compromised Pasteur when he praised him, in the newspaper *Les Mondes,* for having brought "unbelievers and materialists" into the spiritual fold and for having "become aware of his mission," realizing that he was "responsible for souls."

Catholic orators, including Father Félix in the Cathedral of Notre-Dame, congratulated Pasteur on the orthodoxy of his doctrine and stigmatized spontaneous generation as a pernicious, impious, monstrous, and blatantly atheistic theory, spawned by the devil himself.

As a natural result of all this, liberals, anticlericalists, and rationalists vehemently attacked Pasteur. "Honors are heaped on Pasteur because he is orthodox; abuse is heaped on Pouchet, Musset, and Joly because they are intelligent," wrote Edmond About. And Eugène Noël sent this message to Pouchet: "There is no one with a healthy and honest mind, even in the farthest reaches of Asia, who should not take as much interest in your work as do your fellow citizens of Rouen or Toulouse. The freedom of thought of the whole human race is involved. And on what can political and social freedom be established, if not on freedom of thought? It is not only supporters of spontaneous generation who have their eyes on you, it is everyone who wants to preserve the right to think freely."

After these great debates, there were other controversies, some of which proved to be fruitful, between spontanists and antispontanists. Ernest Onimus (1867) thought he saw white corpuscles being born spontaneously in lymph; the chemist Edmond Fremy (1871) affirmed the spontaneous generation of alcoholic yeast; and the Englishman Charlton Bastian (1876), using urine as a putrescible medium, reported new facts that Pasteur was obliged to take into account. It was Bastian's experiments that led to the discovery that heating to a temperature of 100 degrees centigrade was insufficient to destroy germs and that the temperature had to be raised to 120 degrees. This in turn led to the technique of sterilization by steam.

Finally, in 1878, Pasteur again had to defend his ideas, this time against Marceline Berthelot, who referred to a posthumous work by Claude Bernard in which the spontaneous generation of alcoholic yeast was implicitly affirmed.

Pasteur eventually emerged fully victorious from the great struggle. He was able to conclude with confidence: "Spontaneous generation is an illusion. . . . No, there are no known circumstances in which it can be said that microorganisms have come into the world without germs, without parents similar to themselves. Those who deny this have been misled by faulty experiments tainted with errors that they were unable either to see or to avoid."

Pasteur succeeded in solving the problem of spontaneous generation, because, being a chemist experienced in the methods of the exact sciences, he was at least thirty years ahead of his contemporaries, from a technical point of view.

As Etienne Wolff has said, "Pasteur's genius did not lie in having denied spontaneous generation . . . it lay in having found the means of answering the question, having conceived a whole plan of experiments demonstrating that, in nature as it now exists, there is no spontaneous generation of organisms visible under a microscope."

Pasteur's experiments were simple: they followed directly from the techniques inaugurated in the eighteenth century by Needham and Spallanzani; but their correct execution required a clarity of mind, a logical perseverance, an insistence on precision, an ability to sense sources of error in advance, and a tactical ability, so to speak, which belong only to very great researchers.

The latest resurgence of spontanism occurred during the Michurinian delirium that shook Soviet biology in the period around 1948, when there were strange assertions by scientists like Olga Lepeshinskaia and Boshian, who saw microbes being formed in egg yolks, while Trofim Denisovitch Lysenko saw grains of rye appear in ears of wheat.

Pasteur's irrefutable demonstrations were, of course, entirely valid only for the organisms he knew: infusorians, yeasts, microbes. It does not seem impossible that the problem of spontaneous generation may arise again, to some extent, with regard to viruses; but that would be a very different debate, since we do not even know if a virus must be considered a living organism.

The history of ideas on spontaneous generation is a clear, well-defined, "straight-line" history, because it deals with a controversy that differed sharply from many others of its kind, in that truth was not divided between both camps, and the

battle did not end with a compromise between the two opposing views. In this case, we see an outright error stubbornly maintained for several centuries. It was a hardy, tenacious, resilient error; each time it seemed to have been vanquished, it shifted to another area and regained its vigor.

It was an error upheld by great minds: Buffon, Lamarck, Jean-Baptiste Dumas; it found support, at different times, in common sense, logic, experience, and respect for authority; its defenders include freethinkers and political sectarians as well as religious traditionalists.

It was unmasked only gradually, in successive stages. At first, belief in spontaneous generation extended even to mice; then it was limited to snails and lice; then to microbes, infusorians, yeasts. To make the truth emerge, experimentation had to become progressively more thorough, refined, subtle, and precise; methods and techniques had to be perfected; investigators had to become more painstaking.

It cannot be said, however, that the spontanists made no positive contribution. Their stubborn opposition forced their adversaries to make greater efforts to define and clarify the conditions of their experiments; by keeping them under pressure and obliging them to make their position more and more solid, they undoubtedly aided the progress of knowledge.

As Eugène Bataillon says, "The attacks directed against Pasteur's work made him shape and reshape it, giving it a more precise form in response to each objection."

Omne vivum e vivo—No life without preexisting life. It is a truth of immense scope, not only theoretically but also practically, since our medicine, our hygiene, our biology, and even our civilization are to a great extent founded on the anti-spontanist certainty.

It must be clearly understood, however, that although the problem of the present origin of life appears to have been solved without ambiguity, the enormous problem of the first origin of life still remains open. The fact that life is not now formed on our planet from inanimate matter does not necessarily mean that it was not formed in the remote past.

In the present state of our knowledge, there is no longer any belief in a "vital force" added to the elements of matter. There is agreement on the idea that the properties, the attributes of life are manifested when matter acquires a certain degree of ordered complexity, a certain structural arrangement. Some theorists maintain, however, that the material elements already possess a kind of rudimentary life, or "prelife," while others hold that they are totally lacking in life.

These two theories—summation and emergence—bear some relation to the two metaphysical tendencies I described earlier: the Augustinian concept of "seminal spirits" and the Thomistic concept of a "psyche" that appears *de novo* as soon as matter has become capable of receiving it.

Finally, there is the great problem of the artificial creation of life. In whatever way, by whatever process the complexity and structural order capable of giving rise to manifestations of life may have appeared in the past, there is no reason to deny the possibility that human genius may be able to re-create that complexity and order.

This rather Promethean ambition no longer raises any doctrinal objections, even from scientists with a Christian outlook, such as Paul Chauchard, who acknowledges that science may some day discover "the processes by which life came from inanimate matter millions of years ago." It is always advantageous to science when a problem is thus stripped of passion and "dephilosophized."

Personally, I do not think life will be created in a laboratory in the near future, but that is a purely intuitive opinion, and men who know much more about biochemistry than I are more optimistic on the subject.

Thus some day not too far from now, perhaps, man will create life; we may hear the great news on radio or television, between a political platitude and a popular song. If so, it will be quite simply the most important, significant, dramatic, and deeply moving event in all human history, far surpassing trips to the moon and other feats of cosmic prowess.

It will have no military or industrial application, it will not influence the stock market, it will enrich no one, it will give no nation an advantage over any other; but it will exalt man. By manufacturing that humble assimilating and self-reproducing particle, by causing life to be born, for the second time, of something other than itself, man will have closed the great, mysterious cycle. A product of life, he will have in turn become a producer of life. He will have come a little closer to the image he formed of God.

ON BIOGENESIS

Before speculating on the origin of life, I would like to make a semantic remark about the use of the word "biogenesis." It is, of course, perfectly acceptable in the sense in which it will be used during this symposium. Jules Carles, in his excellent little book, *Les Origines de la vie* (Presses Universitaires de France, 1950), prefers it to the term "spontaneous generation," because, he says, as a good philosopher, "To have spontaneity, something must at least exist, and I do not see to the spontaneity of what that generation might be attributed, since, by definition, what is thus engendered did not exist at the moment when it supposedly decided spontaneously to be born. Could it be the spontaneity of another living being? Obviously not, because then it would be a matter of manufacture or creation. One might speak of the spontaneity of inanimate matter, but can the

THIS was the inaugural address of the Symposium on Elementary Biological Systems and Biogenesis (Paris, November 1965).

words 'spontaneous' and 'inanimate' really be applied to the same thing?"

Father Carles finally adopts the word "biogenesis" to designate any birth of life that does not take place in the usual way: when one living organism comes from another—that is, when there is a continuation, a continuity of life—it is formed by generation. If it does not come from another living organism—that is, if there is an absolute beginning, a birth of life—it is produced by biogenesis (assuming that such an event could actually take place).

But to avoid any possibility of misunderstanding, it is important to note that Thomas Huxley, the famous disciple of Darwin, used the word "biogenesis" with a meaning directly opposed to the meaning it will be given in this symposium. For Huxley, biogenesis is the hypothesis that living matter is always produced by the action of previously existing living matter, whereas abiogenesis is the hypothesis that living matter can arise from nonliving matter; that is, Huxley's abiogenesis is what we have here agreed to call biogenesis. If we bear this in mind, there will be no difficulty.

To discuss the origin of life—biogenesis—it would be good to know what life is, but biology must unfortunately declare itself unable to answer that question, though this does not prevent it from increasing our understanding of organic structure and function and steadily giving us greater control of the phenomena of life.

It is worth noting that in the past, in the eighteenth and nineteenth centuries, when there was great ignorance of the manifestations of life and very little control of them, definitions of life were readily given. Here are a few of those definitions, mostly truisms or paradoxes.

KANT: "Life is an inner principle of action." BUFFON: "Life is a Minotaur; it devours the organism." BICHAT: "Life is the aggregate of functions that resist death." Dugès: "Life is the special activity of organized beings." Claude Bernard: "Life is creation." Spencer: "Life is the specific combination of heterogenous changes that are both simultaneous and successive." Blainville: "Life is a double internal movement of decomposition, both general and continuous." Béclard: "Life is organization in action." The *Encyclopédie:* "Life is the opposite of death." And so on. The list could be continued indefinitely, without advancing our understanding in any way.

We have fortunately given up such sterile verbal games and concentrated, more modestly (and, already, not without difficulty), on characterizing the phenomenon of life by its essential attributes: first, the faculty of assimilation, and, following from this, the ability to grow and reproduce.

Perhaps it is good to point out at the start that no one today, or hardly anyone, would think of explaining the manifestations of life by the action of a specific principle, distinct from matter: vital force, psyche, organic soul, entelechy. All biologists, I believe, whatever their philosophical opinions or even their religious beliefs may be, now agree—and this unanimity is worth stressing—that the phenomenon of life is bound up with a certain extremely complex structural arrangement of matter, a certain mode of organization; when that arrangement and that organization are present, what we call the attributes of life are manifested. The problem of life is thus essentially a problem of form and structure. It is in this sense that we can say, with Kahane, that "life does not exist."

The difference of opinion among biologists—or at least among those who allow themselves to philosophize a little—lies in this: some maintain that the form necessary to the first appearance of life must have been produced solely by the action of natural forces, while others believe that it must have had a supernatural origin, or must at least have resulted from the action of certain mysterious energies.

According to Lecomte de Nouÿ, for example, it is impossible to assume that a living molecule could have been formed by pure chance, by random chemical combinations, because such an improbable event would have a reasonable chance of occurring only in a span of time so great that its number of years would be expressed by a figure with 253 digits. The improbability would be so enormous that it would have to be called an impossibility.

Quite impartially, I do not think that such considerations and speciously precise calculations are capable of shedding any light on the problem, because the fact is that we have no way of arriving at even a crude approximation of the odds against the chance formation of a living molecule at the time when life began. We know nothing about conditions on our planet two billion years ago: the state of matter at that time, and so on. Perhaps the formation of life was much more probable than we would dare to conclude from extrapolating present data into the past. Perhaps matter had properties that it has now lost, properties related to a certain state of the cosmos, corresponding to a certain stage in the expansion of the universe. In any case, all our clever reasoning and calculations may very well miss the essential point.

Another divergence among theorists of biogenesis concerns the way in which the properties of life derive from the material architecture that conditions them.

Most biologists believe that the elements of matter are entirely lacking in life and that consequently the property "life" emerges when a certain mode of organization is manifested. Some, however, are inclined to think that these elements are already endowed with a kind of life ("prelife" or "infralife"), so that life, properly speaking, amounts to an effect of summation.

It is interesting for the historian of science to note that this is a very old concept, related to hylozoism.

Already in the eighteenth century, Diderot denied that life and thought could arise from the arrangement of matter. The passage in which he expresses his conviction is worth quoting in its entirety: "It seems to me a rank absurdity, if ever there was one, to suppose that the system of a living body can be formed by placing one, two, or three dead particles* beside another dead particle. What! Are we to believe that when particle A is placed to the left of particle B it is not conscious of its existence, does not feel, is inert and dead, but that when it is placed to the right of particle B they both form a whole which lives, feels, and knows itself? No, that cannot be. What difference do right and left make here? Are there two sides in space? Even if there were, feeling and life would not depend on them. Whatever has those properties has always had them and always will have them."

The idea of the essential vitality of matter came from the geometer Maupertuis, who attributed a principle of intelligence, an elementary form of psychic life (aversion, desire, memory), to the simplest "corpuscles." Like extension, movement, and weight, this principle was one of the properties of matter. He explained the formation of the fetus by a kind of "memory" in the particles of the parental seeds.

"The formation of an organized body," he wrote, "will never be explained solely by the physical properties of matter. How could organization, which is only an arrangement of parts, ever give rise to a feeling?"

On this point, Kant did justice to Maupertuis, who is too often forgotten, by citing him among the restorers of the ancient hylozoism: "Hylozoism animates everything, whereas materialism kills everything. Maupertuis attributes the smallest possible degree of life to the vital organic particles of

---

* In Diderot's time the word "particle" was used instead of "atom."

animals. Other philosophers see these particles only as lifeless masses, serving only to strengthen the levers of the animal machine."

Paul Janet, however, fails to mention Maupertuis among the few materialists more profound than the others (Diderot and Pierre Jean Georges Cabanis, for example) who, "clearly seeing the impossibility of making thought come from what does not think, of making it an accident, a resultant of combinations of extension, maintained that thought, in the form of sensitivity, is an essential property, like weight, movement, and impenetrability."

The idea of the vitality of material particles runs through the whole history of philosophy. In one of the books in which he expressed his aggressive transformism and militant monism, the famous naturalist Ernst Haeckel wrote at the end of the last century: "It is our conviction that atoms already possess sensation and will, or rather feeling (*aesthesis*) and effort (*tropesis*), in their simplest form, that is, a universal soul in the most primitive form." And: "All these changes, whether in organic nature or the inorganic world, seem truly comprehensible to us only if we consider atoms not as small masses of dead matter, but as elementary living particles endowed with forces of attraction and repulsion. The pleasure and displeasure, the love and hatred of atoms are only different expressions of that force of attraction and repulsion."

Haeckel felt that theoretical materialism was wrong to deny all sensation to matter and that not even the simplest physical and chemical phenomena could be explained without attributing a soul to atoms. For him, there was no matter without energy and sensitivity, no energy without matter and sensitivity, no sensitivity without matter and energy. Matter, energy, and sensitivity were the three essential, indissolubly connected attributes of the universe.

Other biologists, no less materialistic than Haeckel, have shared his opinion on the infraconsciousness of matter. "It is natural," wrote Félix Le Dantec, in *Le Déterminisme biologique et la personnalité humaine*, "to assume that a certain combination of atoms is endowed with a consciousness that is the resultant of the elementary consciousnesses of its constituent atoms, rather than considering the consciousness of a complex body as resulting from its very construction, by means of elements lacking in consciousness."

This same idea is expressed by various philosophers of the late nineteenth and early twentieth centuries. Alfred Fouillée, for example: "The elements of psychic life must exist in the elements of apparently inert matter. Since the same law of continuity is applicable to both the psychic world and the physical world, we must be thorough in applying the theory of causality to both, in such a way as to derive more highly developed psychic life from more rudimentary psychic life. In our opinion, this method of analysis leads to recognizing the appetitive process (feeling—appetition) as a universal element in which a felt stimulation gives rise to a more or less conscious reaction." (*Mouvement positiviste.*) And: "Would it not be strange to suppose that there is an abyss between inorganic elements and the living organisms that come from them, and that the phenomena of consciousness are suddenly added, as though having fallen from the sky, to movements of absolutely insensitive matter?"

Finally, Louis Bourdeau (*Le Problème de la vie*, 1901): "The sensitivity manifested in the protean substance does not appear in it miraculously, *ex nihilo*. We must acknowledge that those particles of matter (molecules, atoms) do not represent small, inert, dead masses, but active elements endowed with a kind of inferior life. When we see the highest intelligence gradually develop from an apparently unconscious ovum, we no longer find it difficult to believe that this

ovum, thus endowed with a principle of transcendent spirituality, derives it from the elements of which it is composed."

I have given all these quotations to show how old and
commonplace is the idea that attributes a kind of rudimentary
vitality to the constituent elements of matter.

We know what has been done with this idea by the "Teilhardians," who credit their great mentor with having origi-
nated it.

"If it is impossible," writes Paul Chauchard, "to pass
abruptly from inanimate matter to the cell, we must envisage
a whole process of increasing complexity in inanimate matter,
leading to the final step of life. That is precisely Teilhard de
Chardin's concept of 'prelife.' One of those who have done
most to make us understand this question is the Russian
biochemist Aleksandr Ivanovitch Oparin. We must be grate-
ful to Gavaudan for having given us a French translation of
Oparin's excellent book, under the title *L'origine de la vie sur
la terre*. Gavaudan makes an important contribution of his
own, with his preface and comments. A careful reading of this
book, which is related to Kahane's *La vie n'existe pas*, will
show how scientific progress has obliged both materialists and
spiritualists to develop their views. The former minimized the
properties of life in a simplistic mechanism, the latter attributed them to an extraneous vital force. This is a useless hypothesis because today's materialism must recognize that the
specific traits of life come from the complexity of living
matter. Materialism thus arrives scientifically at conclusive
evidence of a teleology of progress inscribed in the powers of
matter. This should logically lead to a complete examination
of the philosophical significance of that teleology."

Needless to say, I do not agree with my friend Paul
Chauchard on this last point, because I do not believe that

the latest progress of biology, as far as biogenesis is concerned, leads scientifically to "conclusive evidence of a teleology" whose philosophical significance should logically appear to us. The teleologists may be right, and they are naturally entitled to think as they please, but they are not entitled to claim that they have conclusive scientific evidence on their side.

In the past, materialists may have gone too far, when they maintained that science rules out teleology, but it is certain that their opponents go much too far today, when they maintain that science demonstrates teleology.

Whatever opinion we may have on the natural origin of life, we cannot, it seems, exclude the possibility of producing it by artificial means, which would at last permit us to say that we understand it, if it is true, as one thinker has said, that we understand only what we are able to reproduce.

With constantly increasing technical resources at his disposal and with the ability to combine at will the effects of the most powerful energies, why should man not succeed in reconstituting the conditions responsible for the initial biogenesis in a laboratory?

Even those who believe in a transcendent origin of life have no grounds for denying that man, aided by his inventive genius, may eventually succeed in doing something that inanimate matter, left to itself, would have been incapable of doing. In short, if man creates life, he will be imitating ancient nature, according to some, and according to others, he will be imitating the Creator.

Thus Chauchard, a Christian biologist, denies that the artificial creation of life, the dream of all biochemists, would give support to atheism or offend well-informed believers. "If this happens," he says, "it will mean that science has discovered either the process by which life arose from inanimate matter billions of years ago or some other possibility in the properties of matter. When he discovers a secret of nature, the religious scientist will pay homage to the Creator responsible for those properties and for his scientific intelligence."

The problem of the artificial creation of life therefore does not, or at any rate should not, stir up any passions; the synthesis of a living particle cannot prove or disprove any philosophical doctrine.

Is it necessary to recall here that, despite everything said about it in the mass media, the artificial creation of life is not yet in sight? At most, certain constituents of living matter have been synthesized under rather ordinary conditions, and encouraging progress has already been made in this area.

When we speak of creating life, we do not, of course, mean manufacturing a cell, which, with its nucleus, chromosomes, and nucleic acids, is already a whole organized world; we are referring, much more modestly, to a large molecule capable of increasing its mass at the expense of its environment, something rather similar to a virus, except that it will be *auto-trophic*, that is, it will not have to be a parasite of a cell in order to live and reproduce itself, whereas all viruses known to us are incapable of multiplying by themselves and must be reproduced by the cell that harbors them.

Insofar as these viruses are incapable of autonomous life, we may doubt whether they deserve to be considered alive. The same question may be raised with regard to certain organelles found in the normal cell. And our uncertainty in answering it shows the difficulty we have, I will not say in defining life, but in even characterizing it satisfactorily.

Are the chloroplasts of plant cells alive or not? Separated from the cell, they can give off oxygen under the action of light, that is, they can perform the function of chlorophyll

(Hill, 1937). Luigi Califano therefore does not hesitate to say that they are alive, because "what is alive is that which transforms energy, either by taking it from the external world or by liberating it from organic or inorganic compounds."

Tomorrow, perhaps, other organelles will be preserved or cultivated outside the cell and will exhibit more or less vital properties in an artificial environment. Even then, however, there will be room for doubt as to whether their "vitality" is genuine if, in the medium of preservation or culture, they are supplied with substances prepared by living organisms.

Be that as it may, Gavaudan confidently reminds us, in the remarkable comments he has written on Oparin's book, of the progress already made by biochemistry "in the preservation of certain properties of isolated organelles, whether mitochondria or chloroplasts. Can we say that we have reached the end of experimental improvements? The success of such research often depends only on a growth factor or a fortunate choice of initial material. It is obvious that the multiplication *in vitro* of cellular organelles such as mitochondria would be a revolution. Shall we be less bold than Meissner and Hauser, the two researchers cited by Yves Delage . . . who tried to cultivate 'bioblasts' with the artless enthusiasm of innovators who care nothing about difficulties? Is that not the direction taken for a time by Jean Louis Brachet and his colleagues, when they tried to prove by various means that the elementary particles of cytoplasm could be extracted from the cell and cultivated on the chorioallantoic membrane of a chicken? As we know, those experiments unfortunately led only to a mere presumption in favor of the progagation of 'microsomes' by 'division.' " (*L'origine de la vie sur la terre*, Masson, 1965.)

Who has not thought of the possibility of making genes reproduce *in vitro?* Lucien Cuénot and I referred to it as long ago as 1936: "It is permissible to think that some day a way will be found to cultivate genes as we now cultivate cells. 'If we could constantly maintain an environment identical to the one that the action of the neighboring parts continually creates for a given elementary organism, that organism would live in freedom exactly as it does in society.' This sentence, written by Claude Bernard with regard to the cell three-quarters of a century ago, could now be written with regard to the gene." (Lucien Cuénot and Jean Rostand, *Introduction à la Génétique*, Centre de Documentation universitaire, 1936.)

More recently, in his *The Chemistry of Heredity*, Stephen Zamenhof envisages a synthesis of DNA that would bring about "a true reproduction of naked genes in a test tube, a fantastic, perhaps frightening achievement."

If we sometimes feel uncertain as to the life or nonlife of such constituent elements of the cell as chloroplasts, microsomes, and so on, we feel equally uncertain even in the case of a whole cell, after it has been stripped of some of its functions or properties.

I would like to dwell on this point a little because it brings me back to an area that is more familiar to me than biogenesis in general. Let us take a well-known cell as an example: the spermatozoon of the frog. The biologist-philosopher Le Dantec—who often used a rather special terminology—would have said that the spermatozoon is not alive, or at least that it is not endowed with "elementary life," since it is incapable of assimilating and reproducing by itself. "Elements capable of bringing about reproduction are incapable of reproduction." (*Traité de Biologie*.)

But let us pass over that somewhat specious objection.

We can practice a kind of vivisection on the sperm cell—even subtler than the kind devised by Claude Bernard, with poisons or anesthetics—which enables us to dissociate certain properties usually associated with life. We can, for example, remove the spermatozoon's power of fertilization without altering its motility to the slightest degree by treating it with certain substances, notably diluted glycerin.

Can we rightly attribute life to a spermatozoon that can no longer fertilize, can no longer penetrate the ovum, but can still move actively?

Concerning the spermatozoon of the sea urchin, Boris Rybak has written, "I must first make one general remark. It is noteworthy that most authors have judged the life of the spermatozoon by its *motility*, but I now insist that although motility is the criterion of a certain vitality, it is not the criterion of life." (*Histoire de la spermiologie des Oursins*, Biologie Médicale, December 1955.)

Let us pursue our analysis. If frog spermatozoa are treated by various physical agents (X-rays, radium radiation, ultraviolet rays) or chemicals (acraflavine hydrochloride, blue toluidine dye, etc.), they preserve their normal ability to penetrate the ovum and cause the first mitosis of development, but they completely lose their genetic competence; in other words, since their DNA, their genes, have lost the power of self-reproduction, they are unable to contribute in any way to the development of the ovum, which takes place without amphimixis, in what is known as the *gynogenetic* mode, and produces a purely maternal organism.

Such as spermatozoon—motile, capable of penetrating the ovum and initiating its development, but incapable of cooperating in that development—can no doubt be said to be alive; but what are we to say of a spermatozoon which is not motile and cannot penetrate the ovum by itself, but which, if it is introduced artificially with a slender surgical probe, is as capable as a normal, intact sprematozoon of causing the first mitosis of development?

That is exactly the case with a spermatozoon that has been dessicated, frozen in a high vacuum, treated with diluted glycerin for two months, or heated to a temperature of 45 degrees centigrade.

It is generally accepted—especially after Shaver's research on frog eggs—that the power to initiate mitosis is, for the most part, related to mitochondria, the largest of the cytoplasmic microsomes, which are rich in various enzymes, especially respiratory enzymes, and play an important part in protein synthesis.

Since these microsomes have the power to initiate mitosis, are they alive? Or do they only have "a certain vitality," to use Rybak's term? One thing is certain: we know of no inorganic substance, hormone, or enzyme that can be substituted for them.

If we refuse to attribute any vitality to these microsomes, we must grant that substances endowed with the power to initiate mitosis can be isolated in a nonliving state; and since substances endowed with genetic competence can also be isolated in a nonliving state, in the form of DNA, we are led to the conclusion that the two most essential properties of the spermatozoon—genetic competence and the power to initiate mitosis—are independent of life.*

It is interesting to note that study of the power to initiate mitosis is much less advanced than study of the hereditary material.

Since Oskar Hertwig, it has been usual in natural fertilization to dissociate the genetic effect from the effect of stimulation produced on the ovum by the spermatozoon. It might have been assumed—and for a time it actually was assumed, especially after the success of chemical fertilization—that the latter effect would easily be reduced to physiochemical fac-

---

* Cf. J.-André Thomas: "We must distinguish degrees of life, that is, degrees of organization and activity in matter, just as we distinguish degrees of death, and, as we shall see in this book, degrees of survival." (*Survie et conservation biologique*, Masson, 1963.)

tors, whereas the former, being more "vital," would defy physiochemical analysis for a long time, perhaps forever.

But what has happened? We are now beginning to be well informed about the molecular constitution of DNA, which carries genetic information, and we are still quite ignorant with regard to the chemical basis of the power to initiate mitosis. It is as if, contrary to our original assumptions, the power to initiate mitosis were more "vital" than the genetic effect. This is a paradox that deserves to be stressed by a historian of science.

In view of the observations we have just made concerning the independence, the possible dissociation, of the various properties of the frog spermatozoon (motility, ability to penetrate, power to initiate mitosis, genetic capacity), it seems to me that we must agree with Gavaudan when, rejecting the hope of "finding a specifically characteristic property of life, whether it be self-reproduction or reaction to stimuli," he writes, "Life has no unequivical definition. There is no uniformity among the different manifestations that we observe and describe as 'manifestations of life' in a chicken or a mouse, in an isolated cell from a chicken or mouse embryo, in a tree, in a leaf that has been separated from the tree for several months, in a man driving a car, or in a man in a deep coma. The states of life of an isolated heart, a fibroblast in a culture medium, a spermatozoon, a nucleate or non-nucleate red blood corpuscle, a mitochondrion oxidizing fatty acids *in vitro*, and a virulent body infecting a cell, are manifested by very different phenomena which sometimes have no common denominator." (*L'origine de la vie sur la terre.*)

To return to the artificial creation of life, I believe that we are not far—and perhaps we have already done so to a certain extent—from achieving a synthesis of factors or substances endowed with more or less vital properties. Tomorrow, no doubt, we shall manufacture whatever it is in the frog spermatozoon that causes division of the ovum; we shall manufacture whatever it is in frog chromosomes that transmits this or that trait. Will it then be said that we have created a little of the frog's life?

I do not know. The object of this symposium will be to try to answer that question. It may be a purely verbal question, or it may be essential.

ON THE HISTORY
OF SCIENCE

In a time when current events press in upon us from all sides, when the achievements of science and technology urge us to look toward the future rather than turning back to the past, when preparations for interplanetary travel are already under way, and when we soon expect to see artificial hearts, synthesized viruses, and improvements in the human brain, can it reasonably be assumed that there is still strong interest in the history of science?

The answer to that question is given today, I believe, by the success of this congress, which has more than eight hundred participants from all over the world and has received no fewer than five hundred papers.

No one contests the intellectual importance of the history of science. It unquestionably forms a major chapter in the history of thought and civilization. George Sarton has said that a

THIS was the inaugural address of the Twelfth International Congress on the History of Science (Paris, August 26, 1968.)

genuine humanist must know the life of science as he knows the life of art and religion.

What can be more instructive, richer in subjects of reflection for the psychologist, the epistemologist, and the philosopher, than to follow the slow and laborious embryogeny of truth through the ages?

By what means, what procedures, does man gain the intellectual maturity that enables him to form an increasingly refined representation of the real world and thereby extend his power to act on it? Are we to assume, with Jean Piaget, that the cognitive functions are a mental extension of the organic regulatory functions, so that it is biology's task to clarify the mechanisms set in motion by the acquisition of knowledge? What are the personal or collective conditions which favor that acquisition? What is the role of technology, produced by science but capable of giving it strong assistance in return? How do the different disciplines influence each other, and how does science as a whole benefit from the emancipation that it owes to philosophical reflection? Does the evolution of scientific thought always move "in the direction of history?" Shall we acknowledge, with Michel Foucault, that analysis of the progress of the sciences, particularly biology, can be an accomplice of that "destruction of man" that has recently been announced to us, not without a certain secret satisfaction?

Independently of its speculative scope, the history of science has a purely esthetic value. When we look back on the construction of the majestic cathedral of truths and errors that constitutes the knowledge of a given period, we see "a beautiful spectacle that delights the mind," with increasingly numerous artisans constantly correcting and reshaping it, or adding new elements to its structure.

It is also a highly moral spectacle. " 'History of science,' " says Boris Rybak. "I cannot help being moved by those words: they are the equivalent of 'history of the mind.' "

They give us the stirring notion of the continuity of the human achievement and the solidarity of minds through time and space, since each truth has international roots that extend far back into the past. Whether he likes it or not, every scientist is a "citizen of the world."

The history of science also shows us the dignity of even the humblest effort, for there are no brilliant discoveries that have not been preceded and prepared by obscure endeavors. What Edmond Rostand said about great military leaders can be said about all great discoverers: "And you would be nothing without the dark, humble army that it takes to compose a page of history."

And finally, a lesson of confidence and modesty emerges from the history of science. Confidence in the powers of the intellect, since there is something in everyone's work that ultimately honors mankind; but also modesty, since we see how fallible the workers are, and how, even in the best minds, truth has difficulty in triumphing over preconceptions, vanity, doctrinal obstinacy, and "epistemological obstacles," as Gaston Bachelard has said.

Because of the wealth and diversity of its content, it has never been decided, in France at least, whether the teaching of the history of science belongs to science or to the humanities.

Although no one doubts the educational and humanistic value of the history of science, there is disagreement about the profit that the research scientist can draw from it. Some have gone so far as to say that the spirit of erudition, turned toward the past, is contrary to the spirit of research, turned toward the future; that the kind of curiosity that leads one to the direct study of nature is incompatible with the kind that leads one to a library.

This was the opinion of François Magendie and also Maximilien Littré, who denied that science had roots in earlier ages. The great Claude Bernard considered it futile to exhume faulty observations or discredited theories and feared that science might regress if young men were oriented toward the study of old books. He even felt that knowledge of modern scientific work should not be carried too far, lest it dry out the mind, smother imagination, and impair originality.

The philosopher Yvon Belaval holds a similar view today: "Is all culture based on historical studies? No. The mathematician or the scientist has no need to concern himself with the past of his discipline, because for him it is dead and therefore useless. He learns what can now be accomplished in his field. These accomplishments, though new, will soon be surpassed, and will then become historical to the historian of science. They take place in the present indicative of the timeless, so to speak, like the enunciation of a theorem or a law."

At the opposite end of the spectrum are the opinions of the chemist Jean-Baptiste Dumas, who doubted that anyone could understand science without going back to its sources. Of Isidore Geoffroy Saint-Hilaire, who saw knowledge of the past as a necessary condition for comprehension of the present. Of Pasteur: "We do not know a problem well unless we have followed its development from the beginning." Of Emile Gley: "Like living organisms, the great scientific questions, particularly those in biology, cannot be clearly understood without knowledge of their evolution from their origins to the present." Of Bachelard: "It seems to me that one cannot understand the atom of modern physics without evoking the history of its imagery." Of Louis de Broglie, finally, who remarks that "each science bears within itself the ineradicable traces of a long hereditary past," and concludes that every scientist should know the past of his discipline, just as an anatomist should know the embryogeny of the structures he studies.

It is undeniable that contemporary scientists differ greatly in the interest they take in the history of science. Many of them see it only as an extraneous pastime, a kind of intellectual luxury. When a problem demands their attention, they approach it as it currently presents itself, in its most advanced state, without worrying about the phases it may have gone through in the past. Recognizing no need to connect today's thought with yesterday's, they feel that their time is well taken up with assimilating knowledge directly useful to them, preparing their experiments, and learning the best techniques. And it can be predicted, I believe, that scientists of this kind, voluntarily enclosed in the present, will become more and more numerous, considering the mass of information they must acquire solely in order to carry out their laboratory work. When a scientist is already overwhelmed by the swelling flood of research that pours in from all countries, when he is goaded by fear of being outdistanced by someone else and thus deprived of credit for a discovery or an invention, how can he even momentarily turn away from fruitful immediate activity to consult old treatises?

There are indeed very few cases in which a researcher has been able to find profitable inspiration in an old scientific work; and however fascinating we may find the history of science, we must admit that most scientific problems can be satisfactorily approached without knowing anything at all about the evolution of ideas that brought them to their present state.

It is even permissible to wonder whether knowledge of the past may not restrict freedom of judgment, since well-founded theories have sometimes been unjustly opposed because they were seen as a return to some ancient belief. This was true of the atomic theory, so fiercely contested by Dumas and then Bertholet, both very erudite men; it was also true of Lamarck's transformism, Pasteur's germ theory, and even the science of genetics in its early stages.

Some scientists rejected the chromosome theory of heredity

because they thought they caught a whiff of "preformism" in it. Le Dantec accused Weismann of conceiving the germ as "the nonfigurative equivalent of an invisible homunculus," and Morgan pertinently pointed out that the superficial similarity between the gene theory and theories of representative particles gave opponents of Mendelism an opportunity to attack it. Only recently, the embryologist Paul Wintrebert assailed the notions of the gene and of organizers in embryos because he judged them to be tainted with preformism. And the term "chemical homunculus" has been applied to the nucleic acid molecules that carry the hereditary information.

For my part, I believe that, for a scientist, the main value of the history of science lies in the emotional stimulus it provides. I therefore agree with Etienne Wolff (who is himself a great researcher and a man who has meditated deeply on the paths to discovery) when he writes, "I cannot too strongly recommend reading, and making students read, the lives and works of scientists. Along with textbooks and standard treatises, which are a bit arid and sometimes even grim, one should also read living scientific works. Unfortunately they are rare. The best are those written by scientists themselves. Such works are capable of giving young people a taste for research and of providing science with new recruits; in short, they can create vocations."

Among these stimulating and inciting books, I will cite—and I apologize for citing mostly French works, but it is because of ignorance, not partiality—Claude Bernard's *Introduction à l'étude de la médecine expérimentale*, Charles Darwin's autobiography, Charles Richet's *Souvenirs d'un physiologiste*, Charles Nicolle's *Biologie de l'Invention*, Jean-Henri Fabre's *Souvenirs entomologiques*, Bataillon's *Enquête sur la génération*, and Etienne Wolff's *Les Chemins de la vie*.

And to this list I will add a few books written about scientists by scientists, or by laymen respectful of their subjects: *Pierre Curie,* by Marie Curie; *Pasteur,* by René Dubos; *La Vie de Pasteur,* by René Vallery-Radot; *L'Histoire d'un esprit,* by Emile Duclaux; *Theodor Schwann,* by Marcel Florkin; and *Fleming,* by André Maurois.

Such books help to animate and humanize science: they imbue static, inert knowledge with the vitality of living truth. Through them, we sense a kind of emotion that is expressed in no other category of writing; we feel the grandeur, even the poetry, of the struggle with the unknown—and sometimes with oneself—that constitutes the adventure of research. We share the disappointments and victories of the researcher, we accompany him in his gropings; we identify with those singular men who place above all other rewards the satisfaction of possessing a truth that has never appeared to anyone else.

The mind ignites the mind. When a young man, attracted to science, hears Claude Bernard speak of the "unsurpassable joys" to be found in the pursuit of truth; when he hears Pasteur confess the kind of intoxication that came over him when he had verified his hypothesis on the dimorphism of crystals with a polarimeter; when he hears Madame Curie tell of the never-failing "rapture" she felt whenever she saw the first radioactive preparations glowing in the darkness of the famous shed; and when he hears Jean Eugène Bataillon describe the sleepless nights that followed the discovery of traumatic parthenogenesis, how can he fail to dream of the time when he, too, may know those joys, intoxications, raptures, and sleepless nights?

Contrary to what the physiologist Paul Bert said, it is not quite true that, although science develops the intelligence, "only the humanities make the heart beat," by giving to thought "that sublime disinterestedness which makes us learn and reflect for the pure satisfaction of knowing."

When he set up that opposition, Bert overlooked the values

peculiar to the history of science. According to the classic view, only the "we" reigns in science; by restoring the "I" to it, the history of science changes it from a mere collection of facts into a gallery of faces and a succession of personal efforts. By identifying with this or that scientist of the past, a young man may strengthen a vocation, an ambition.

According to the psychoanalyst Raymond de Saussure, young Sigmund Freud identified with certain "giants" of science, notably Darwin. This "adolescent megalomania," stemming from narcissism, may ultimately benefit the human race. Hence "the energizing value of biography," to quote André Maurois, an expert on the subject. "Nothing has more influence on men's acts than knowledge of other men's acts. The reader tries to resemble the hero and act like him. Imitation of a great man is one form of moral education."

One may object, with Father Russo, to a concept of the history of science that attributes a little too much value to the part played by a few great men regarded as "heroes" and "giants of the mind," but at least it thus gives honorable idols to those who have a need to admire and respect.

It is good for us to know even the weaknesses and inadequacies of those great figures of the past—dead men who are sometimes more alive than the living—because they are instructive and encouraging. A young apprentice naturalist may benefit from knowing that the author of *The Origin of Species* was a poor pupil, that he considered himself to be of only average intelligence, and that in his childhood he had a tendency to tell false stories.

I would also like to stress the special value of certain very old books, which, although they are no longer of any real scientific interest, can still play an important part in the development of a researcher's sensitivity and "superego."

It seems to me that something would be lacking in a naturalist who had never read the pages in which Réaumur depicts the hatching of a dragonfly, Charles Bonnet reports placing an aphid under glass to make it procreate in solitude, Abraham Trembley explains how he discovered the regenerative power of the freshwater polyp, and Lazzaro Spallanzani describes the experiments that led him to practice the artificial insemination of toads.

How can anyone fail to be struck by the magnificent simplicity of those observations and discoveries, which kindle the imagination all the more because they were carried out with no technical apparatus, solely by the powers of ingenuity, skill, and patience?

Those old authors are the youth of science, its childhood, and are therefore closer to the apprentice researcher. I would compare them to the "primitives" in art.

"Those whom we call ancient," said Pascal, "were actually new in all things, and formed the childhood of man in the proper sense of the word." When Pascal expressed himself in that way, it was to excuse the errors and ignorance of the ancients. But if I call them "children," it is to stress their freshness of mind and their marvelous spontaneity, from which we, who are fatigued by too much knowledge, can derive understanding and stimulation.

Thus, across the centuries, we form friendships based on a brotherhood of interest and inclination. Is he not our remote brother, that man who contemplated the same organisms that we have before our eyes, who asked questions about them so close to the ones that concern us today? And would it not be ingratitude on our part if we ignored the gift he gave us by showing us the way?

There is often talk of the doctrinal lessons that can be drawn from the history of science. Although I would not go so far as to say, with Paul Valéry, that history—"the most dangerous product ever fashioned by the chemistry of the intellect"—is incapable of teaching us anything, because "it contains everything and gives examples of everything," I will admit that the teachings of the history of science should be used only with circumspection.

It is undoubtedly instructive to learn from the history of science that truth is seldom wholly on one side and that most of the great theoretical quarrels have ended with a compromise between the two antagonistic concepts. This was the case with the opposition between epigenesis and preformation, between the exogenous and the endogenous origins of disease, between the cellular and the humoral theories of immunity, and so on. Every truth is, of course, incomplete and provisional and can be accepted only subject to alteration. Think of all the errors we recited in our youth!

But I do not think that this is any justification for taking refuge, whenever a debate arises, in a kind of lazy "neutralism," in which the mind loses some of its buoyancy and eagerness for combat. I have known biologists of that kind who, foreseeing a future reconciliation, hesitated to choose between orthodox genetics and the theory propounded by Lysenko; and I now know some who are waiting for a reconciliation between Lamarckism and Neo-Darwinism.

Science, it seems to me, has nothing to gain from such refusals of commitment. Let us not forget that there are, after all, quarrels in which one of the adversaries is completely right; and even when truth is not entirely on one side, there is usually one side that works toward it better than the other.

The history of science is also used rather often to combat a certain dogmatism that rejects the possibility of this or that reality because it is contrary to the orthodoxy of the moment. There is a long list of "follies" that were later recognized as truths: official science refused to believe that stones could fall from the sky; on hearing the first phonograph, it suspected the hidden presence of a ventriloquist; it jeered at Boucher de

Perthes when he affirmed the existence of fossil men, at Freud when he deciphered the language of the unconscious, at Karl von Frisch when he deciphered the language of bees.

Resistance to subversive truths forms a chapter in the history of science that is one of the longest and by no means the least colorful. A famous nonconformist, Auguste Lumière, wrote a whole book on the subject: *Les Fossoyeurs du Progrès*.

But here, too, it is good to keep a sense of proportion. We must not allow the gross blunders of true science to be exploited by advocates of false sciences—occultists, astrologers, magicians—who, by constantly reminding us of them, try to make us feel uneasy about our critical rigor when we reject their nonsense.

Every judgment has its risk; only a timorous mind will refrain from judging for fear of error. It is the same here as in art, in which fear of overlooking an "unknown beauty," as Anatole France said, should not make us indulgent toward all kinds of inanity and derangement.

If there is one notion that clearly emerges from the history of science, and from which we can learn something, it is, I believe, the extreme diversity of the personal qualities and abilities that have contributed to the advancement of our knowledge.

What disparities there are among scientists, what variety in their aptitudes, tastes, inclinations, intellectual styles, sensitivities, temperaments!

We find both logical and intuitive thinkers among them, experimenters and theorists, collectors of small facts and creators of vast systems; they may be either bold or cautious, skillful or clumsy, rigorous or haphazard, patient or impatient, docile or rebellious, conventionally educated or self-taught, scholarly or untutored, doubting or assertive. We find men of analysis and men of synthesis, men who dream and men who scrutinize, men who guess and men who demonstrate.

From this we can draw a valuable scholastic conclusion, namely, that we should make higher education available to some of the young people who are now excluded from it because they do not fit certain predetermined norms, because they are not cast in a certain intellectual mold.

Without wishing to reopen here a debate whose importance I have often stressed, I will take as a specific example the false but still persistent idea that all biological research requires the use of mathematics. It would be easy to show, by the history of science, that many if not most discoveries in biology have been made without the aid of mathematics and by means of qualities and aptitudes that are given little or no consideration in the recruitment of scientists.

The famous geneticist Bridges had an exquisitely "sharp eye" that enabled him to distinguish the subtlest mutations of color in the eye of a fruit fly. Morgan made him an associate in his research without asking him whether he knew how to solve an equation.

One scientist may owe the success of his work to a rare dexterity, another may owe it to his inexhaustible patience or the strong motivation that led him into research. Let us recall that Darwin regarded his ardent, steadfast love of the natural sciences as one of the foremost reasons for his own success. I am inclined to agree with Jean Sénebier when he says that a man is made for truth if he is capable of seeking it with passion.

In short, if we make a selection among candidates for research, as I assume we must, it should be made in several different directions, as widely varied as possible, to avoid impoverishing our pool of intellectual abilities and creating "pure strains," which is always contrary to fecundity.

By our present methods of recruitment, not only do we create many discontented people who will always be embittered by their exclusion from the work that would have best suited their inclinations, but we also undoubtedly deprive science (and I am thinking especially of biology) of good minds capable of doing fruitful research.

Do we really have so many seekers—and finders—of scientific truth that we are justified in discouraging all those eager aspirants, simply because they do not have the qualities demanded by official educators?

Consider how literature and the arts would be impoverished if selection in those fields were as biased as it is in the sciences.

By showing us the extreme diversity of the factors involved in scientific creativity, the history of science teaches us that we should open the doors of our laboratories more widely. If we put that lesson into practice, our reflection on the past will have had a beneficial effect on the future.